NEWS

NASA

NATIONAL AERONAUTICS AND SPACE ADMINISTRATION
WASHINGTON, D.C. 20546

TELS. WO 2-4155
WO 3-6925

FOR RELEASE: THURSDAY A.M.
April 2, 1970

RELEASE NO: 70-50K

PROJECT: APOLLO 13

I0040864

P
R
E
S
S

K
I
T

contents

-more-

Cover: Damaged Apollo 13
This view of the severely damaged Apollo 13 service module was photographed from the lunar module after it was jettisoned. The command module, still docked with the Lunar Module, is in the foreground. An entire panel on the service module was blown away by the apparent explosion of oxygen tank number two located in Sector 4. Three fuel cells, two oxygen tanks and two hydrogen tanks were located in Sector 4. The damaged area is forward (above) the S-Band high gain antenna. The damage to the module caused the Apollo 13 crew members to use the lunar module as a lifeboat. The lunar module was jettisoned just prior to Earth re-entry. The Apollo 13 crew consisted of mission commander James A. Lovell, command module pilot John L. Swigert and lunar module pilot Fred W. Haise Jr.
Image Credit: NASA

Published by Books Express Publishing
Copyright © Books Express, 2012
ISBN 978-1-78039-862-4

Books Express publications are available from all good retail and online booksellers. For publishing proposals and direct ordering please contact us at: info@books-express.com

FOR RELEASE: THURSDAY A.M.
April 2, 1970

RELEASE NO: 70-50

APOLLO 13 THIRD LUNAR LANDING MISSION

Apollo 13, the third U.S. manned lunar landing mission, will be launched April 11 from Kennedy Space Center, Fla., to explore a hilly upland region of the Moon and bring back rocks perhaps five billion years old.

The Apollo 13 lunar module will stay on the Moon more than 33 hours and the landing crew will leave the spacecraft twice to emplace scientific experiments on the lunar surface and to continue geological investigations. The Apollo 13 landing site is in the Fra Mauro uplands; the two National Aeronautics and Space Administration previous landings were in mare or "sea" areas, Apollo 11 in the Sea of Tranquility and Apollo 12 in the Ocean of Storms.

Apollo 13 crewmen are commander James A. Lovell, Jr.; command module pilot Thomas K. Mattingly III, and lunar module pilot Fred W. Haise, Jr. Lovell is a U.S. Navy captain, Mattingly a Navy lieutenant commander, and Haise a civilian.

-more-

3/26/70

Launch vehicle is a Saturn V.

Apollo 13 objectives are:

*   Perform selenological inspection, survey and sampling of materials in a preselected region of the Fra Mauro formation.

*   Deploy and activate an Apollo Lunar Surface Experiment Package (ALSEP).

*   Develop man's capability to work in the lunar environment.

*   Obtain photographs of candidate exploration sites.

Currently 11 television transmissions in color are scheduled: one in Earth orbit an hour and a half after launch, three on the outward voyage to the Moon; one of the landing site from about nine miles up; two from the lunar surface while the astronauts work outside the spacecraft; one at the command service module/lunar module docking operation; one of the Moon from lunar orbit; and two on the return trip.

The Apollo 13 landing site is in the hilly uplands to the north of the crater Fra Mauro. Lunar coordinates for the landing site are 3.6 degrees south latitude by 17.5 degrees west longitude, about 95.6 nautical miles east of the Apollo 12 landing point at Surveyor III crater.

Experiments emplaced at the Fra Mauro site as part of the ALSEP III will gather and relay long-term scientific data to Earth for at least a year on the Moon's physical and environmental properties. Five experiments are contained in the ALSEP: a lunar passive seismometer will measure and relay meteoroid impacts and moonquakes; a heat flow experiment will measure the heat flux from the lunar interior to the surface and conductivity of the surface materials to a depth of about 10 feet; a charged particle lunar environment experiment will measure protons and electrons to determine the effect of the solar wind on the lunar environment; a cold cathode gauge experiment will measure density and temperature variations in the lunar atmosphere; and a dust detector experiment.

The empty third stage of the Saturn V launch vehicle will be targeted to strike the Moon before the lunar landing and its impact will be recorded by the seismometer left by the Apollo 12 astronauts last November. The spent lunar module ascent stage, as in Apollo 12, will be directed to impact the Moon after rendezvous and final LM separation to provide a signal to both seismometers.

Candidate future Apollo landing sites -- Censorinus, Davy Rille, and Descartes -- will be photographed with a large-format lunar topographic camera carried for the first time on Apollo 13. The lunar topographic camera will make high-resolution 4.5 inch square black-and-white photos in overlapping sequence for mosaics or in single frames. The camera mounts in the command module crew access hatch window when in use. After lunar orbit rendezvous with the lunar module and LM jettison the command module will make a plane-change maneuver to drive the orbital track over Descartes and Davy Rille for topographic photography.

The Apollo 13 flight profile in general follows those flown by Apollos 11 and 12 with one major exception: lunar orbit insertion burn no. 2 has been combined with descent orbit insertion and the docked spacecraft will be placed into a 7x57 nautical mile lunar orbit by use of the service propulsion system. Lunar module descent propellant is conserved by combining these maneuvers to provide 15 seconds of additional hover time during the landing.

Lunar surface touchdown is scheduled to take place at 9:55 p.m. EST April 15, and two periods of extravehicular activity are planned at 2:13 a.m. EST April 16 and 9:58 p.m. EST April 16. The LM ascent stage will lift off at 7:22 a.m. April 17 to rejoin the orbiting command module after more than 33 hours on the lunar surface.

Apollo 13 will leave lunar orbit at 1:42 p.m. EST April 18 for return to Earth. Splashdown in the mid-Pacific just south of the Equator will be at 3:17 p.m. EST April 21.

After the spacecraft has landed, the crew will put on clean coveralls and filter masks passed in to them through the hatch by a swimmer, and then transfer by helicopter to a Mobile Quarantine Facility (MQF) on the USS Iwo Jima. The MQF and crew will be offloaded in Hawaii and placed aboard a C-141 aircraft for the flight back to the Lunar Receiving Laboratory at the Manned Spacecraft Center in Houston. The crew will remain in quarantine up to 21 days from completion of the second EVA.

The crew of Apollo 13 selected the call signs Odyssey for the command/service module and Aquarius for the lunar module. When all three crewmen are aboard the command module, the call sign will be "Apollo 13." As in the two previous lunar landing missions, an American flag will be emplaced on the lunar surface. A plaque bearing the date of the Apollo 13 landing and the crew signatures is attached to the LM.

Apollo 13 backup crewmen are USN commander John W. Young, commander; civilian John L. Swigert, Jr., command module pilot; and USAF Major Charles M. Duke, Jr., lunar module pilot.

APOLLO 13 - LAUNCH TO LUNAR SURFACE

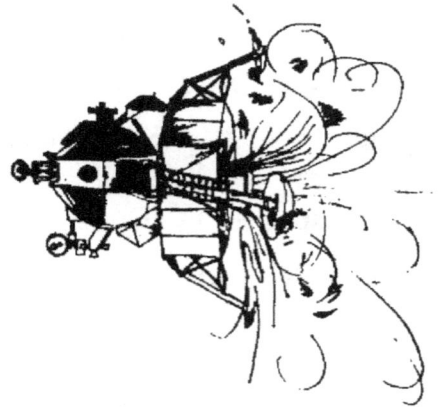

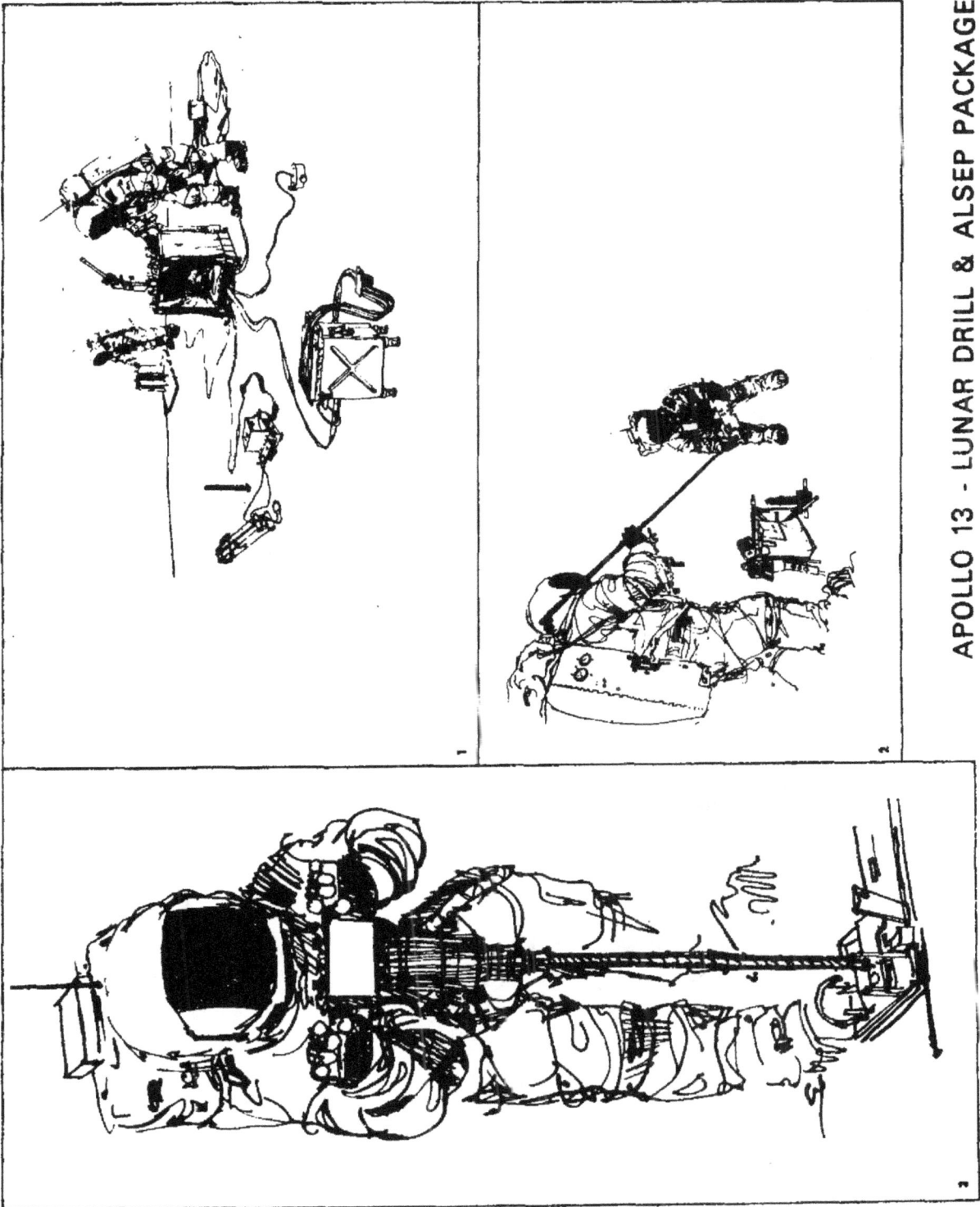

APOLLO 13 - LUNAR DRILL & ALSEP PACKAGE

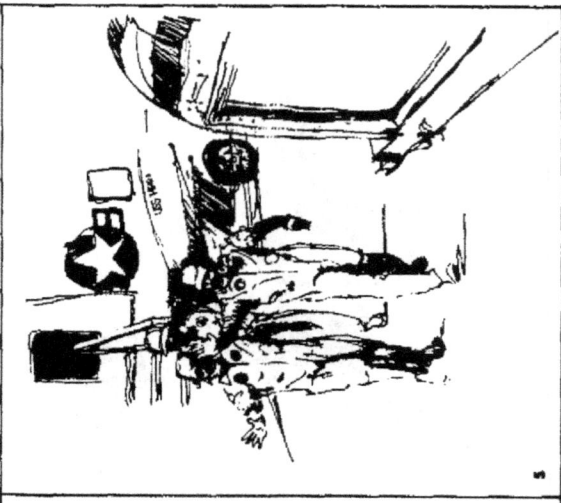

APOLLO 13 - LUNAR SURFACE TO RECOVERY

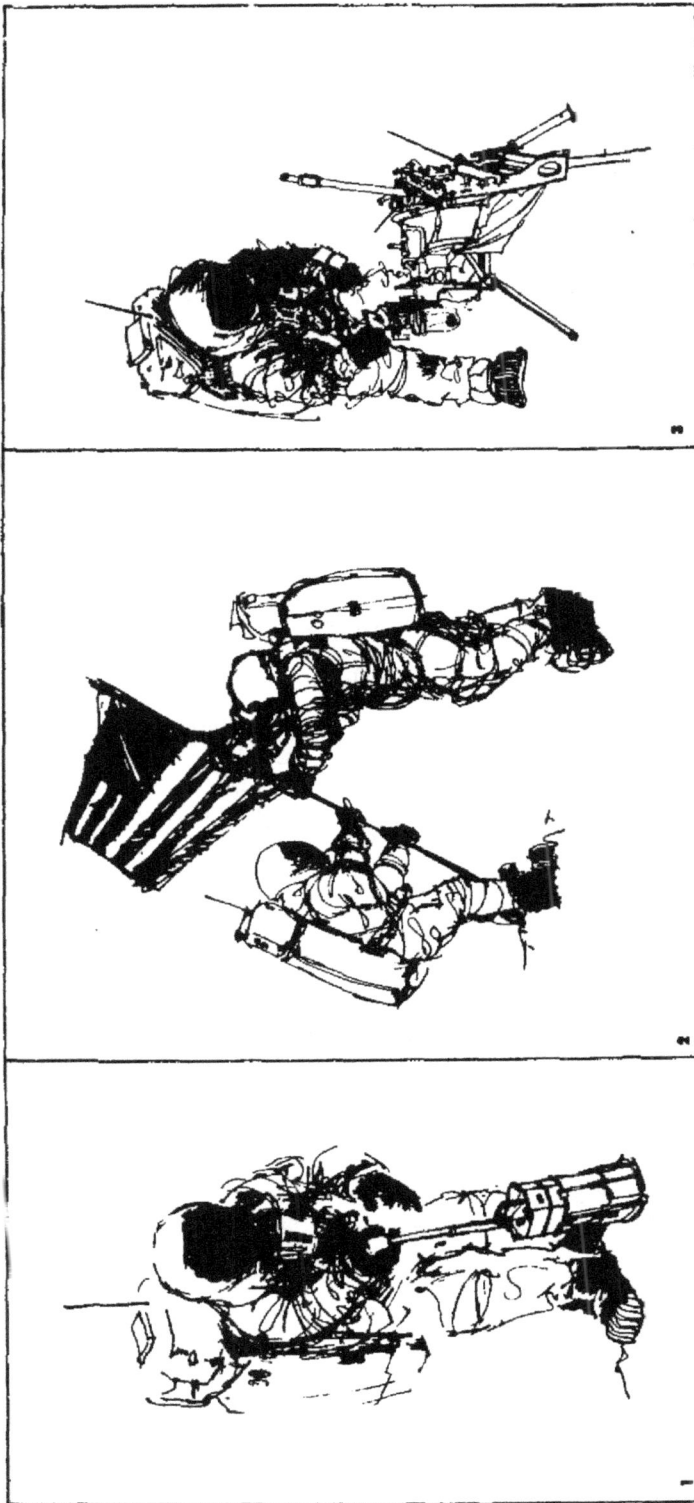

APOLLO − 13 − LUNAR SURFACE ACTIVITIES

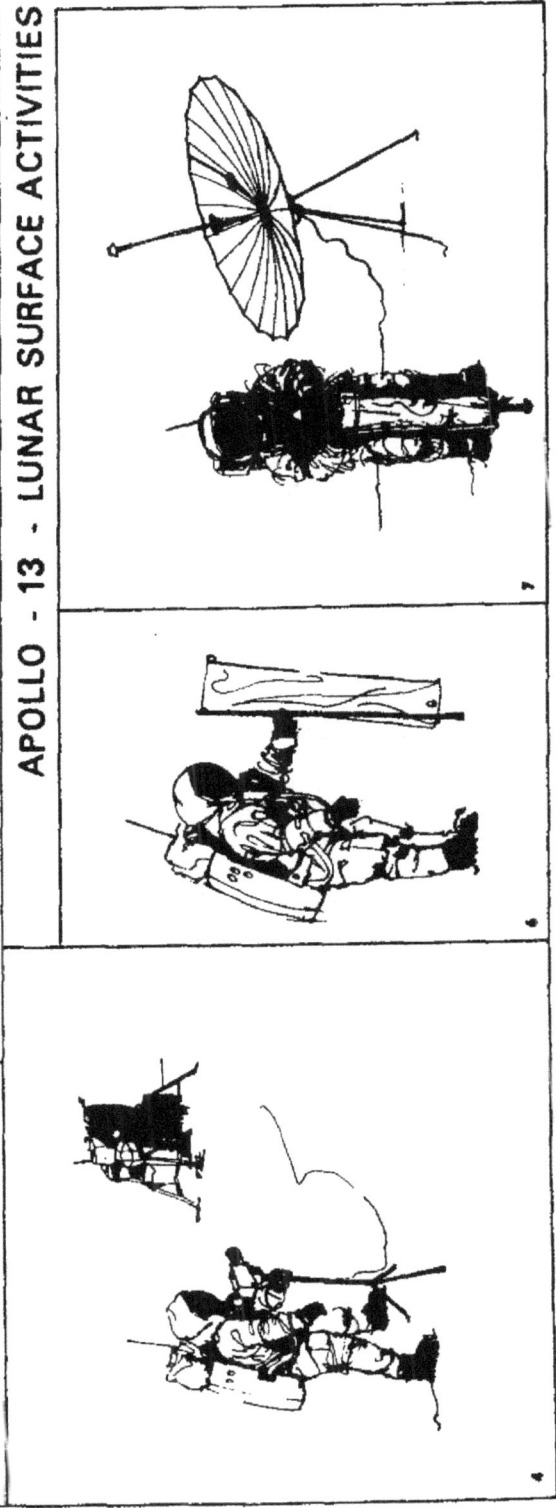

APOLLO 13 COUNTDOWN

Precount activities for the Apollo 13 launch begin about T-4 days, when the space vehicle will be prepared for the start of the official countdown. During precount, final space vehicle ordnance installation and electrical connections will be accomplished. Spacecraft gaseous oxygen and gaseous helium systems will be serviced, spacecraft batteries will be installed, and LM and CSM mechanical buildup will be completed. The CSM fuel cells will be activated and CSM cryogenics (liquid oxygen - liquid hydrogen) will be loaded and pressurized.

The countdown for Apollo 13 will begin at T-28 hours and will continue to T-9 hours, at which time a built-in hold is planned prior to the start of launch vehicle propellant loading.

Following are some of the major operations in the final count:

T-28 hours	Official countdown starts LM stowage and cabin closeout (T-31:30 to T-18:00)
T-27 hours, 30 minutes	Install and connect LV flight batteries (to T-23 hours)
T-22 hours, 30 minutes	Topoff of LM super critical helium (to T-20 hours, 30 minutes)
T-19 hours, 30 minutes	LM SHe thermal shield installation (to T-15 hours, 30 minutes) CSM crew stowage (T-19 to T-12 hours, 30 minutes)
T-16 hours	LV range safety checks (to T-15 hours)
T-15 hours	Installation of ALSEP FCA to T-14 hours, 45 minutes)
T-11 hours, 30 minutes	Connect LV safe and arm devices (to 10 hours, 45 minutes) CSM pre-ingress operations (to T-8 hours 45 minutes)
T-10 hours, 15 minutes	Start MSS move to parksite

T-9 hours	Built-in hold for 9 hours and 13 minutes. At end of hold, pad is cleared for LV propellant loading
T-8 hours, 05 minutes	Launch vehicle propellant loading - Three stages (LOX in first stage, LOX and LH_2 in second and third stages). Continues thru T-3 hours 38 minutes
T-4 hours, 17 minutes	Flight crew alerted
T-4 hours, 02 minutes	Medical examination
T-3 hours, 32 minutes	Breakfast
T-3 hours, 30 minutes	One-hour hold
T-3 hours, 07 minutes	Depart Manned Spacecraft Operations Building for LC-39 via crew transfer van.
T-2 hours, 55 minutes	Arrive at LC-39
T-2 hours, 40 minutes	Start flight crew ingress
T-2 hours	Mission Control Center - Houston/ spacecraft command checks
T-1 hour, 55 minutes	Abort advisory system checks
T-1 hour, 51 minutes	Space Vehicle Emergency Detection System (EDS) test
T-43 minutes	Retract Apollo access arm to stand-by position (12 degrees)
T-42 minutes	Arm launch escape system
T-40 minutes	Final launch vehicle range safety checks (to 35 minutes)
T-30 minutes	Launch vehicle power transfer test LM switch over to internal power
T-20 minutes to T-10 minutes	Shutdown LM operational instrumentation
T-15 minutes	Spacecraft to internal power

T-6 minutes	Space vehicle final status checks
T-5 minutes, 30 seconds	Arm destruct system
T-5 minutes	Apollo access arm fully retracted
T-3 minutes, 7 seconds	Firing command (automatic sequence)
T-50 seconds	Launch vehicle transfer to internal power
T-8.9 seconds	Ignition sequence start
T-2 seconds	All engines running
T-0	Liftoff

Note: Some changes in the above countdown are possible as a result of experience gained in the countdown demonstration test which occurs about 10 days before launch.

Lightning Precautions

During the Apollo 12 mission the space vehicle was subjected to two distinct electrical discharge events. However, no serious damage occurred and the mission proceeded to a successful conclusion. Intensive investigation led to the conclusion that no hardware changes were necessary to protect the space vehicle from similar events. For Apollo 13 the mission rules have been revised to reduce the probability that the space vehicle will be launched into cloud formations that contain conditions conducive to initiating similar electrical discharges although flight into all clouds is not precluded.

May Launch Opportunities

The three opportunities established for May -- in case the launch is postponed from April 11 -- provide, in effect, the flexibility of a choice of two launch attempts. The optimum May launch window occurs on May 10. The three day window permits a choice of attempting a launch 24 hours earlier than the optimum window and if necessary a further choice of a 24 hour or 48 hour recycle. It also permits a choice of making the first launch attempt on the optimum day with a 24-hour recycle capability. The May 9 window (T-24 hrs) requires an additional 24 hours in lunar orbit before initiating powered descent to arrive at the landing site at the same time and hence have the same Sun angle for landing as on May 10. Should the May 9 window launch attempt be scrubbed, a decision will be made at that time, based on the reason for the scrub, status of spacecraft cryogenics and weather predictions, whether to recycle for May 10 (T-0 hrs) or May 11 (T+24 hrs). If launched on May 11, the flight plan will be similar for the May 10 mission but the Sun elevation angle at lunar landing will be 18.5° instead of 7.8°.

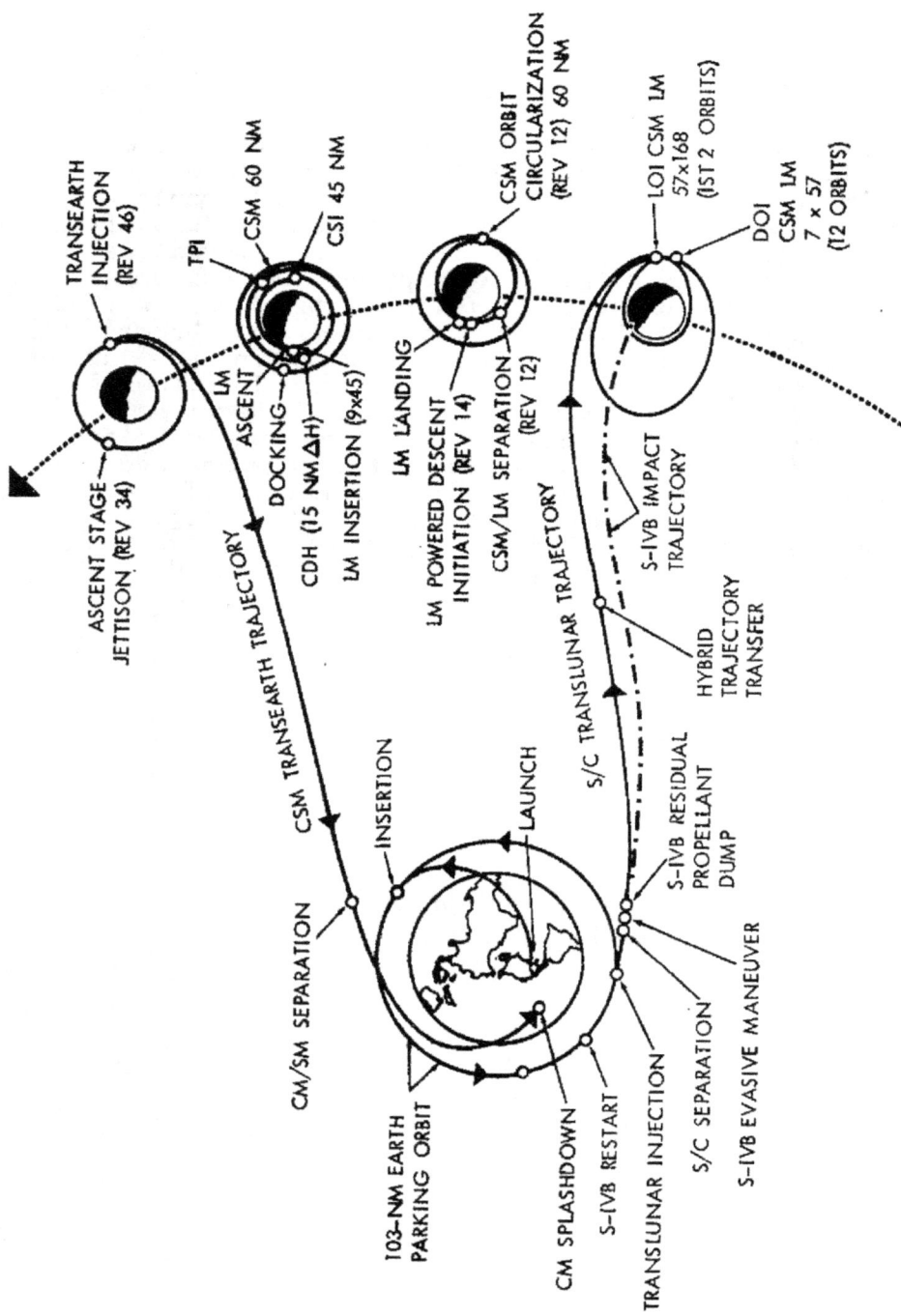

APOLLO 13 FLIGHT PROFILE

LAUNCH, MISSION TRAJECTORY AND MANEUVER DESCRIPTION

The information presented here is based on an on-time April 11 launch and is subject to change before or during the mission to meet changing conditions.

Launch

A Saturn V launch vehicle will lift the Apollo 13 spacecraft from Launch Complex 39A, NASA-Kennedy Space Center, Fla. The azimuth may vary from 72 to 96 degrees, depending on the time of launch. The azimuth changes with launch time to permit a fuel-optimum injection from Earth parking orbit to a free-return circumlunar trajectory.

April 11 launch plans call for liftoff at 2:13 p.m. EST on an azimuth of 72 degrees. The vehicle will reach an altitude of 36 nautical miles before first stage cutoff 51 nm downrange. During the 2 minutes 44 seconds of powered flight, the first stage will increase vehicle velocity to 7,775 feet per second.* First stage thrust will reach a maximum of 8,995,108 pounds before center engine cutoff. After engine shutdown and separation from the second stage, the booster will fall into the Atlantic Ocean about 364 nm downrange from the launch site (30 degrees North latitude and 74 degrees West longitude) about 9 minutes 4 seconds after liftoff.

The second stage (S-II) will carry the space vehicle to an altitude of 102 nm and a distance of 892 nm downrange. At engine shutdown, the vehicle will be moving at a velocity of 21,508 fps. The four outer J-2 engines will burn 6 minutes 32 seconds during the powered phase, but the center engine will be cut off 4 minutes 47 seconds after S-II ignition.

At outboard engine cutoff, the S-II will separate and, following a ballistic trajectory, plunge into the Atlantic about 2,450 nm downrange from the Kennedy Space Center (31 degrees North latitude and 33.4 degrees West longitude) some 20 minutes 41 seconds after liftoff.

The single engine of the Saturn V third stage (S-IVB) will ignite about 3 seconds after the S-II stage separates. The engine will fire for 143 seconds to insert the space vehicle into a circular Earth parking orbit of 103 nm beginning about 1,468 nm downrange. Velocity at Earth orbital insertion will be 24,243 fps at 11 minutes 55 second ground elapsed time (GET). Inclination will be 33 degrees to the equator.

*NOTE: Multiply nautical miles by 1.1508 to obtain statute miles; multiply feet per second by 0.6818 to obtain statute miles per hour.

The crew will have a backup to launch vehicle guidance during powered flight. If the Saturn instrument unit inertial platform fails, the crew can switch guidance to the command module systems for first-stage powered flight automatic control. Second and third stage backup guidance is through manual takeover in which spacecraft commander hand controller inputs are fed through the command module computer to the Saturn instrument unit.

Launch Events

Time Hrs.	Min.	Sec.	Event	Altitude Feet	Velocity Ft/Sec*	Range Naut. Mi.
00	00	00	First Motion	198	0	0
00	01	23	Maximum Dynamic Pressure	42,139	1,600	3
00	02	15	S-IC Center Engine Cutoff	139,856	5,120	24
00	02	44	S-IC Outboard Engines Cutoff	218,277	7,775	51
00	02	45	S-IC/S-II Separation	220,576	7,804	52
00	02	46	S-II Ignition	225,368	7,788	54
00	03	14	S-II Aft Interstage Jettison	300,222	8,173	80
00	03	20	LET Jettison	313,619	8,276	95
00	07	43	S-II Center Engine Cutoff	588,840	17,650	603
00	09	18	S-II Outboard Engines Cutoff	615,508	21,503	892
00	09	19	S-II/S-IVB Separation	615,789	21,517	895
00	09	22	S-IVB Ignition	616,616	21,518	906
00	11	45	S-IVB First Cutoff	627,996	24,239	1,429
00	11	55	Parking Orbit Insertion	628,014	24,243	1,468

* Not including velocity due to Earth's rotation, about 1,350 feet-per-second.

Apollo 13 Mission Events

Events	GET Hrs.:Min.	Date/EST	Vel. Change Feet/Sec.	Purpose & Resultant Orbit
Earth orbit insertion	00:11	11 2:24 p.m.	25,593	Insertion into 103 nm circular Earth parking orbit.
Translunar injection (S-IVB Engine ignition)	02:35	11 4:48 p.m.	10,437	Injection into free-return translunar trajectory with 210 nm pericynthion.
CSM separation, docking	03:06	11 5:19 p.m.	--	Hard-mating of CSM and LM
Ejection from SLA	04:00	11 6:14 p.m.	1	Separates CSM-LM from S-IVB-SLA.
S-IVB evasive maneuver	04:19	11 6:32 p.m.	9.4	Provides separation prior to S-IVB propellant dump and thruster maneuver to cause lunar impact.
Midcourse correction 1	TLI+9 hrs	12 1:54 a.m.	*0	*These midcourse corrections have a nominal velocity change of 0 fps, but will be calculated in real time to correct TLI dispersions. MCC-2 is an SPS maneuver (15 fps) to lower pericynthion to 59 nm; trajectory then becomes non-free return.
Midcourse correction 2 (Hybrid transfer)	30:41	12 8:54 p.m.	15	
Midcourse correction 3	LOI-22 hrs.	13 9:38 p.m.	*0	
Midcourse correction 4	LOI- 5 hrs.	14 2:38 p.m.	*0	
Lunar orbit insertion	77:25	14 7:38 p.m.	-2.815	Inserts Apollo 13 into 57x168 nm elliptical lunar orbit.
S-IVB impact	77:46	14 7:59 p.m.	--	Seismic event.

Events	GET Hrs.:Min.	Date/EST	Vel. Change Feet/Sec.	Purpose & Resultant Orbit
Descent orbit insertion	81:45	14 11:58 p.m.	-213	SPS burn places CSM/LM into 7x57 nm lunar orbit.
CSM-LM undocking	99:16	15 5:29 p.m.	--	Establishes equiperiod orbit for 2.5 nm separation at PDI maneuver.
CSM circularization	100:35	15 6:48 p.m.	70	Inserts CSM into 52x62 nm orbit.
LM powered descent initiation (PDI)	103:31	15 9:44 p.m.	-6635	Three-phase maneuver to brake LM out of transfer orbit, vertical descent and touchdown on lunar surface.
LM touchdown on lunar surface.	103:42	15 9:55 p.m.		Lunar exploration, deploy ALSEP lunar surface geological sample collection, photography.
Depressurization for first lunar surface EVA.	108:00	16 2:13 a.m.		
CDR steps to surface	108:15	16 2:29 a.m.		
CDR collects contingency samples	108:21	16 2:34 a.m.		
LMP steps to surface	108:27	16 2:40 a.m.		
CDR unstows and erects S-band antenna	108:32	16 2:45 a.m.		
LMP mounts TV camera on tripod	108:34	16 2:47 a.m.		

Events	GET Hrs.:Min.	Date/EST	Vel. Change Feet/Sec.	Purpose & Resultant Orbit
LMP reenters LM to switch to S-Band antenna	108:43	16 2:56 a.m.		
LMP returns to lunar surface	108:57	16 3:10 a.m.		
CDR deploys U.S. flag	109:04	16 3:17 a.m.		
CDR and LMP begin unstowing and deployment of ALSEP	109:30	16 3:43 a.m.		
CDR and LMP return to LM collecting samples enroute	111:10	16 5:23 a.m.		
CDR and LMP arrive back at LM, stow gear and samples	111:20	16 5:33 a.m.		
LMP deploys solar wind composition experiment	111:34	16 5:47 a.m.		
LMP reenters LM	111:43	16 5:56 a.m.		
CDR reenters LM	111:58	16 6:11 a.m.		
LM hatch closed, repress	111:59	16 6:12 a.m.		
CSM plane change	113:43	16 7:56 a.m.		
Depress for EVA-2	127:45	16 9:58 p.m.		
CDR steps to surface	127:58	16 10:11 p.m.		
LMP steps to surface	128:07	16 10:20 p.m.		

Event	GET Hrs.:Min.	Date/EST	Vel. Change Feet/Sec.	Purpose & Resultant Orbit
Begin field geology travers, collect core tube and gas analysis samples, dig soil mechanics trench, magnetic sample collection.	128:18	16 10:31 p.m.		
Complete geology traverse	131:04	17 1:17 a.m.		
Return to LM area, retrieve solar wind experiment, stow gear and samples.	131:05	17 1:18 a.m.		
LMP enters LM	131:28	17 1:41 a.m.		
CDR transfers samples, LMP assists	131:35	17 1:48 a.m.		
CDR enters LM, close hatch	131:41	17 1:54 a.m.		
Repress cabin	131:44	17 1:57 a.m.		
LM ascent	137:09	17 7:22 a.m.	6,044	Boosts stage into 9x45 nm lunar orbit for rendezvous with CSM.
Insertion into lunar orbit.	137:16	17 7:29 a.m.		
LM RCS concentric sequence initiation (CSI) burn	138:19	17 8:32 a.m.	50	Raises LM perilune to 44 nm, adjusts orbital shape for rendezvous sequence.
LM RCS constant delta height (CDH) burn	139:04	17 9:17 a.m.		Radially downward burn adjusts LM orbit to constant 15 nm below CSM.

Event	GET Hrs.:Min.	Date/EST	Vel. Change Feet/Sec.	Purpose & Resultant Orbit
LM RCS terminal phase	139:46	17 9:59 a.m.	24.7	LM thrusts along line of sight toward CSM, mid-course and braking maneuvers as necessary.
Rendezvous (TPF)	140:27	17 10:40 a.m.		Completes rendezvous sequence.
Docking	140:45	17 10:58 a.m.		Commander and LM pilot transfer back to CSM.
LM jettison, separation (SM RCS)	143:04	17 1:17 p.m.		Prevents recontact of CSM with LM ascent stage during remainder of lunar orbit.
LM ascent stage deorbit (RCS)	144:32	17 2:45 p.m.	-186	Seismometer records impact event.
LM ascent stage impact	145:00	17 3:13 p.m.		Impact at about 5,508 fps, at -48 angle 35 nm from Apollo 13 ALSEP.
Plane change for photos	154:13	18 12:26 a.m.		Descartes and Davy-Rille photography.
Transearth injection (TEI) SPS	167:29	18 1:42 p.m.	3,147	Inject CSM into trans-earth trajectory.

Event	GET Hrs.:Min.	Date/EST		Vel. Change Feet/Sec.	Purpose & Resultant Orbit
Midcourse correction 5	182:31	19	4:44 a.m.	0	Transearth midcourse corrections will be computed in real time for entry corridor control and recovery area weather avoidance.
Midcourse correction 6	EI-22 hrs	20	5:03 p.m.	0	
Midcourse correction 7	EI- 3 hrs	21	12:03 p.m.	0	
CM/SM separation	240:34	21	2:47 p.m.	--	Command module oriented for entry.
Entry interface(400,000 feet)	240:50	21	3:03 p.m.	--	Command Module enters Earth's sensible atmosphere at 36,129 fps.
Splashdown	241:04	21	3:17 p.m.	--	Landing 1,250 nm downrange from entry 1.5° South latitude by 157.5° West longitude.

Earth Parking Orbit (EPO)

Apollo 13 will remain in Earth parking orbit for one and one-half revolutions. The final "go" for the TLI burn will be given to the crew through the Carnarvon, Australia, Manned Space Flight Network station.

Translunar Injection (TLI)

Midway through the second revolution in Earth parking orbit, the S-IVB third-stage engine will restart at 2:35 GET over the mid-Pacific Ocean near the equator and burn for almost six minutes to inject Apollo 13 toward the Moon. The velocity will increase from 25,593 fps to 36,030 fps at TLI cutoff to a free return circumlunar trajectory from which midcourse corrections could be made with the SM RCS thrusters.

Transposition, Docking, and Ejection (TD&E)

After the TLI burn, the Apollo 13 crew will separate the command/service module from the spacecraft module adapter (SLA), thrust out away from the S-IVB, turn around and move back in for docking with the lunar module. Docking should take place at about three hours and 21 minutes GET. After the crew confirms all docking latches solidly engaged, they will connect the CSM-to-LM umbilicals and pressurize the LM with oxygen from the command module surge tank. At about 4:00 GET, the spacecraft will be ejected from the spacecraft LM adapter by spring devices at the four LM landing gear "knee" attach points. The ejection springs will impart about one fps velocity to the spacecraft. A 9.4 fps S-IVB attitude thruster evasive maneuver in plane at 4:19 GET will separate the spacecraft to a safe distance from the S-IVB.

Saturn Third Stage Lunar Impact

Through a series of pre-set and ground-commanded operations, the S-IVB stage/instrument unit will be directed to hit the Moon within a target area 375 nautical miles in diameter, centered just east of Lansberg D Crater (3 degrees South latitude; 30 degrees West longitude), approximately 124 miles west of the Apollo 12 landing site.

The planned impact will provide a seismic event for the passive seismometer experiment placed on the lunar surface by the Apollo 12 astronauts in November 1969.

The residual propellants in the S-IVB will be used to attempt the lunar impact. Part of the remaining liquid oxygen (LOX) will be dumped through the engine for 48 seconds to slow the vehicle into a lunar impact trajectory. The liquid hydrogen tank's continuous venting system will vent for five minutes.

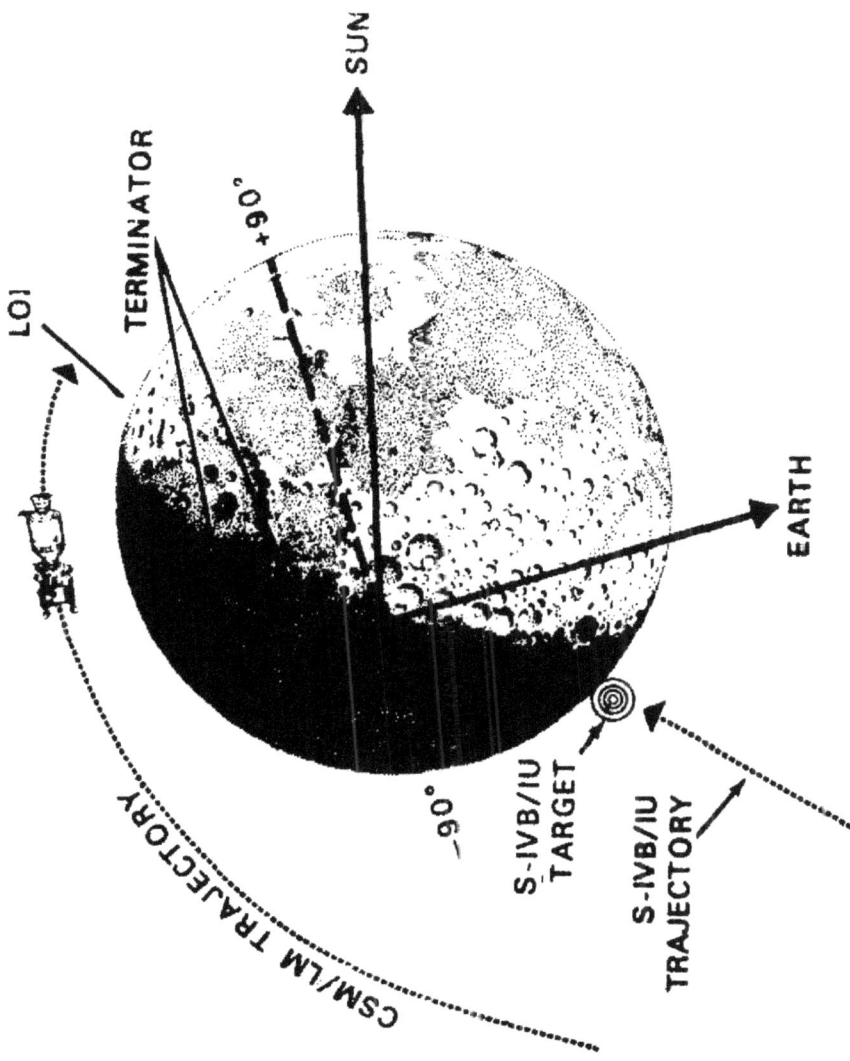

S-IVB LUNAR IMPACT

A mid-course correction will be made with the stage's auxiliary propulsion system (APS) ullage motors. A second APS burn will be used if necessary, at about 9 hours GET, to further adjust the impact point. Burn time and attitude will be determined from onboard systems and tracking data provided to ground controllers by the Manned Space Flight Network.

The LOX dump by itself would provide a lunar impact; the mid-course correction burns will place the S-IVB/IU within the desired target area for impact about 20 minutes after the command/service module enters lunar orbit.

The schedule of events concerning the lunar impact is:

Time Hrs:Min	Event
02 42	Translunar injection (TLI) -- maneuver completion
04 19	Begin S-IVB evasive maneuver (APS engines)
04 21	End evasive maneuver
04 36	LH_2 tank continuous vent on
04 41	Begin LOX dump
04 41	LH_2 tank continuous vent off
04 42	End LOX dump
06 00	Begin first APS burn
08 59	Begin final APS burn (if required)
09 04	APS ullage engines off
77 46	Lunar impact of S-IVB/IU

Translunar Coast

Up to four midcourse correction burns are planned during the spacecraft's translunar coast, depending upon the accuracy of the trajectory resulting from the TLI maneuver. If required, the midcourse correction burns are planned at TLI+9 hours, TLI+ 30 hours, 41 minutes, lunar orbit insertion (LOI)-22 hours and LOI-5 hours. The MCC-2 is a 15 fps SPS hybrid transfer maneuver which lowers pericynthion from 210 nm to 59 nm and places Apollo 13 on a non-free-return trajectory.

Return to the free-return trajectory is always within the capability of the spacecraft service propulsion or descent propulsion systems.

During coast periods between midcourse corrections, the spacecraft will be in the passive thermal control (PTC) or "barbecue" mode in which the spacecraft will rotate slowly about its roll axis to stabilize spacecraft thermal response to the continuous solar exposure.

Lunar Orbit Insertion (LOI)

The lunar orbit insertion burn will be made at 77:25 GET at an altitude of about 85 nm above the Moon. The LOI burn will have a nominal retrograde velocity change of 2815 fps and will insert Apollo 13 into a 57x168 nm elliptical lunar orbit.

Descent Orbit Insertion (DOI)

A 213 fps SPS retrograde burn at 81:45 GET will place the CSM /LM into a 7x57 nm lunar orbit from which the LM will begin the later powered descent to landing. In Apollos 11 and 12, DOI was a separate maneuver using the LM descent engine. The Apollo 13 DOI maneuver in effect is a combination LOI-2 and DOI and produces two benefits: conserves LM descent propellant that would have been used for DOI and makes this propellant available for additional hover time near the surface, and allows 11 lunar revolutions of spacecraft tracking in the descent orbit to enhance position/velocity (state vector) data for updating the LM guidance computer during the descent and landing phase.

Lunar Module Separation

The lunar module will be manned and checked out for undocking and subsequent landing on the lunar surface north of the crater, Fra Mauro. Undocking during the 12th revolution will take place at 99:16 GET. A radially downward service module RCS burn of 1 fps will place the CSM on an equiperiod orbit with a maximum separation of 2.5 nm.

CSM Circularization

During the 12th revolution, a 70 fps posigrade SPS burn at 100:35 GET will place the CSM into 52x62 nm lunar orbit, which because of perturbations of the lunar gravitational potential, should become nearly circular at the time of rendezvous with the LM.

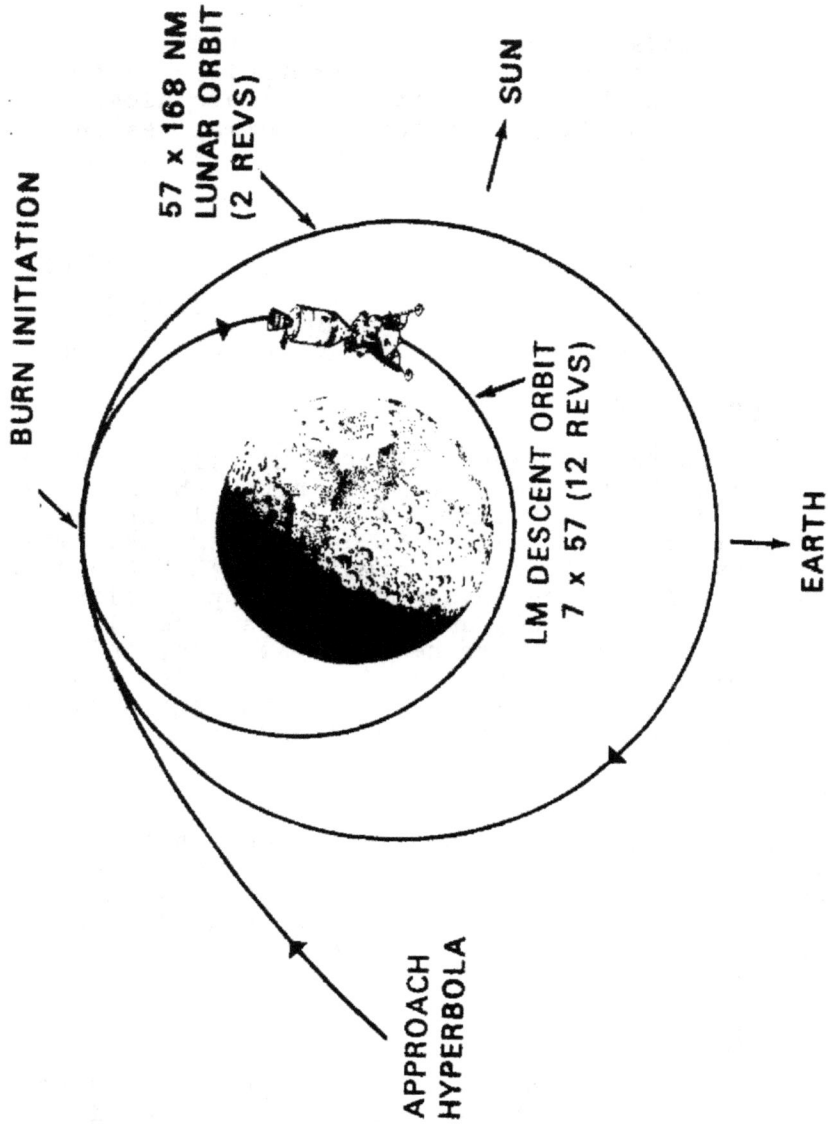

Power Descent Initiation (PDI), Lunar Landing

During the 14th revolution a three-phase powered descent (PD) maneuver begins at pericynthion at 103:31 GET using the LM descent engine to brake the vehicle out of the descent orbit. The guidance-controlled PD maneuver starts about 260 nm prior to touchdown, and is in retrograde attitude to reduce velocity to essentially zero at the time vertical descent begins. Spacecraft attitude will be windows up from powered descent initiation to the end of the braking phase so that the LM landing radar data can be integrated continually by the LM guidance computer and better communications can be maintained. The braking phase ends at about 7,400 feet above the surface and the spacecraft is rotated more toward an upright windows-forward attitude to permit a view of the landing site. The start of the approach phase is called high gate, and the start of the landing phase at about 500 feet is called low gate.

Both the approach (visibility) phase and landing phase allow pilot takeover from guidance control as well as visual evaluation of the landing site. The final vertical descent to touchdown begins at about 100 feet when all forward velocity is nulled out. Vertical descent rate will be 3 fps. The crew may elect to take over manual control at approximately 500 feet. The crew will be able to return to automatic landing control after a period of manned maneuvering if desirable. Touchdown will take place at 103:42 GET.

Lunar Surface Exploration

During the 33 1/2 hours Apollo 13 commander James Lovell and lunar module pilot Fred Haise are on the surface, they will leave the lunar module twice for four-hour EVAs. These are extendable to five hours in real time if the physical conditions of the astronauts and amount of remaining consumables permit.

In addition to gathering more data on the lunar environment and bringing back geological samples from a third lunar landing site, Lovell and Haise will deploy a series of experiments which will relay back to Earth long-term scientific measurements of the Moon's physical and environmental properties.

The experiments series, called the Apollo Lunar Surface Experiment Package (ALSEP), will be left on the surface and could transmit scientific and engineering data to the Manned Space Flight Network for at least a year.

The ALSEP for Apollo 13, stowed in the LM descent stage scientific equipment bay, comprises components for the five ALSEP experiments -- passive seismic, heat flow, charged particle lunar environment, cold cathode gauge, and lunar dust detector.

CSM CIRCULARIZATION (REV 12)
(62 N. MI. BY 52 N. MI.)

LM DESCENT ORBIT
(57 N. MI. BY 7 N. MI.)

UNDOCKING AND
SEPARATION (REV 12)

SUN

EARTH

LANDING
(REV 14)   PDI

LM DESCENT ORBITAL EVENTS

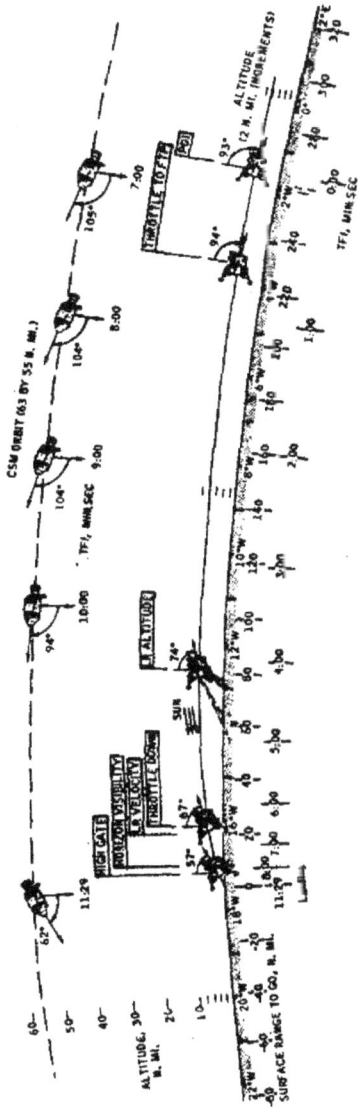

POWERED DESCENT PROFILE

EVENT	SUMMARY				
	TFI, MIN-SEC	V, FPS	$\dot{H}$, FPS	H, FT	ΔV, FPS
POWERED DESCENT INITIATION	0:00	5562	-4	50 700	0
THROTTLE TO MAXIMUM THRUST	0:26	5530	-5	50 611	32
LANDING RADAR ALTITUDE UPDATE	4:12	3180	-84	38 824	2427
THROTTLE RECOVERY	6:30	1466	-82	24 181	4248
LANDING RADAR VELOCITY UPDATE	6:50	3509	-210	23 205	4424
HORIZON VISIBILITY	8:14	639	-183	11 401	5205
HIGH GATE	8:34	478	-185	7 555	5406
LOW GATE	10:16	54 to 77ᵃ	-13	503	6224
LANDING	11:29	-15 to 0ᵃ	-3	5	6633

ᵃHORIZONTAL VELOCITY RELATIVE TO SURFACE

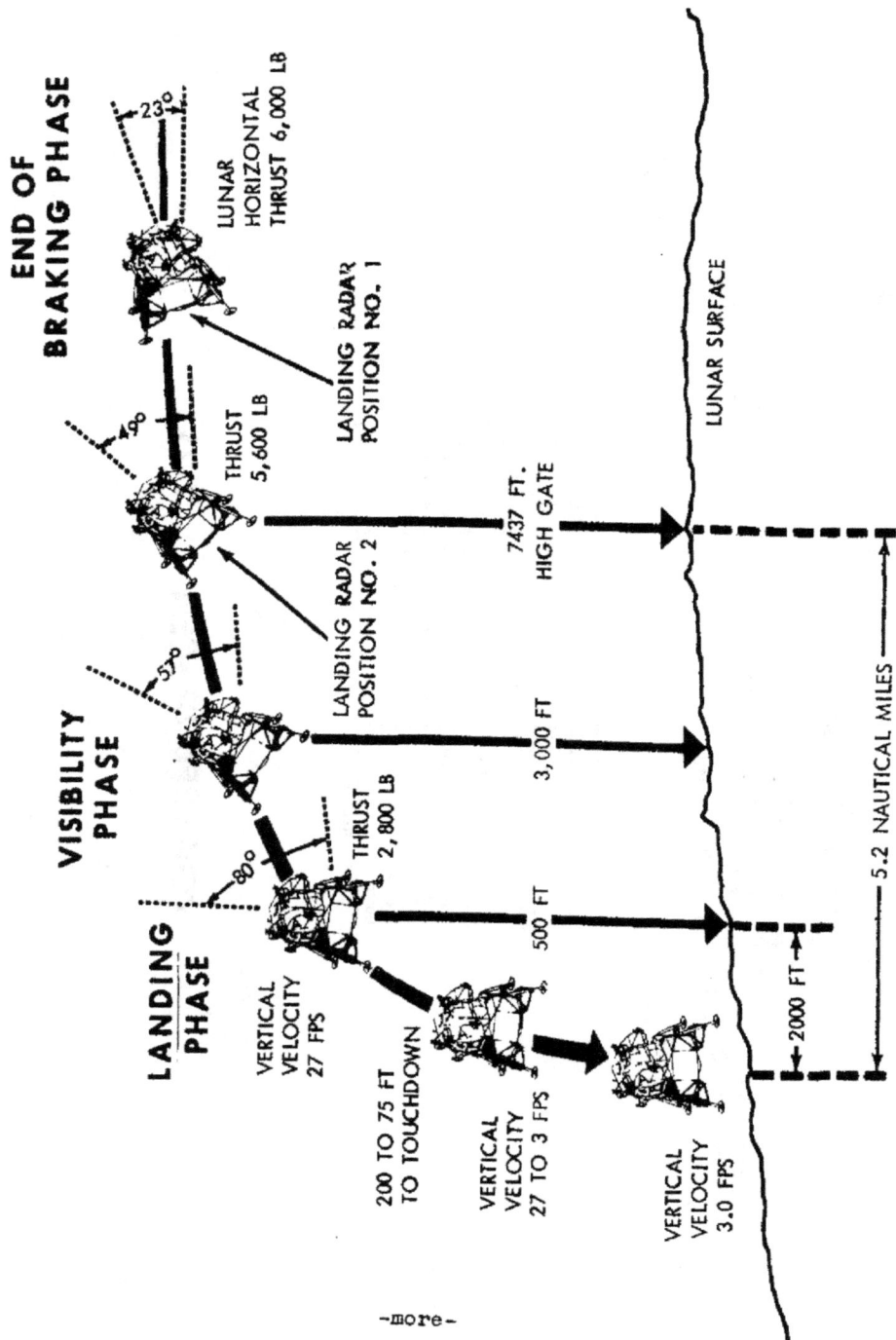

END OF BRAKING PHASE

23°

LUNAR HORIZONTAL THRUST 6,000 LB

LANDING RADAR POSITION NO. 1

49°

THRUST 5,600 LB

LANDING RADAR POSITION NO. 2

7437 FT. HIGH GATE

57°

VISIBILITY PHASE

3,000 FT

LANDING PHASE

80°

THRUST 2,800 LB

500 FT

VERTICAL VELOCITY 27 FPS

200 TO 75 FT TO TOUCHDOWN

VERTICAL VELOCITY 27 TO 3 FPS

VERTICAL VELOCITY 3.0 FPS

LUNAR SURFACE

2000 FT

5.2 NAUTICAL MILES

NOMINAL DESCENT TRAJECTORY FROM HIGH GATE TO TOUCHDOWN

These experiments are aimed toward determining the structure and state of the lunar interior, the composition and structure of the lunar surface and processes which modify the surface, and evolutionary sequence leading to the Moon's present characteristics. The Passive Seismic Experiment will become the second point in a lunar seismic net begun with the first ALSEP at the Surveyor III landing site of Apollo 12. Those two seismometers must continue to operate until the next seismometer is emplaced to complete the three-station set. The heat flow experiment includes drilling two 10-foot holes with the lunar surface drill.

While on the surface, the crew's operating radius will be limited by the range provided by the oxygen purge system (OPS), the reserve backup for each man's portable life support system (PLSS) backpack. The OPS supplies 45 minutes of emergency breathing oxygen and suit pressure.

Among other tasks assigned to Lovell and Haise for the two EVA periods are:

*Collect a contingency sample of about two pounds of lunar material.

*Gather about 95 pounds of representative lunar surface material, including core samples, individual rock samples and fine-grained fragments from the Fra Mauro hilly uplands site. The crew will photograph thoroughly the areas from which samples are taken.

*Make observations and gather data on the mechanical properties and terrain characteristics of the lunar surface and conducting other lunar field geological surveys, including digging a two-foot deep trench for a soil mechanics investigation.

*Photograph with the lunar stereo closeup camera small geological features that would be destroyed in any attempts to gather them for return to Earth.

*Deploy and retrieve a windowshade-like solar wind composition experiment similar to the ones used in Apollos 11 and 12.

Early in the first EVA, Lovell and Haise will set up the erectable S-Band antenna near the LM for relaying voice, TV, and LM telemetry to MSFN stations. After the antenna is deployed, Haise will climb back into the LM to switch from the LM steerable S-Band antenna to the erectable antenna while Lovell makes final adjustments to the antenna's alignment. Haise will then rejoin Lovell on the lunar surface to set up a United States flag and continue with EVA tasks.

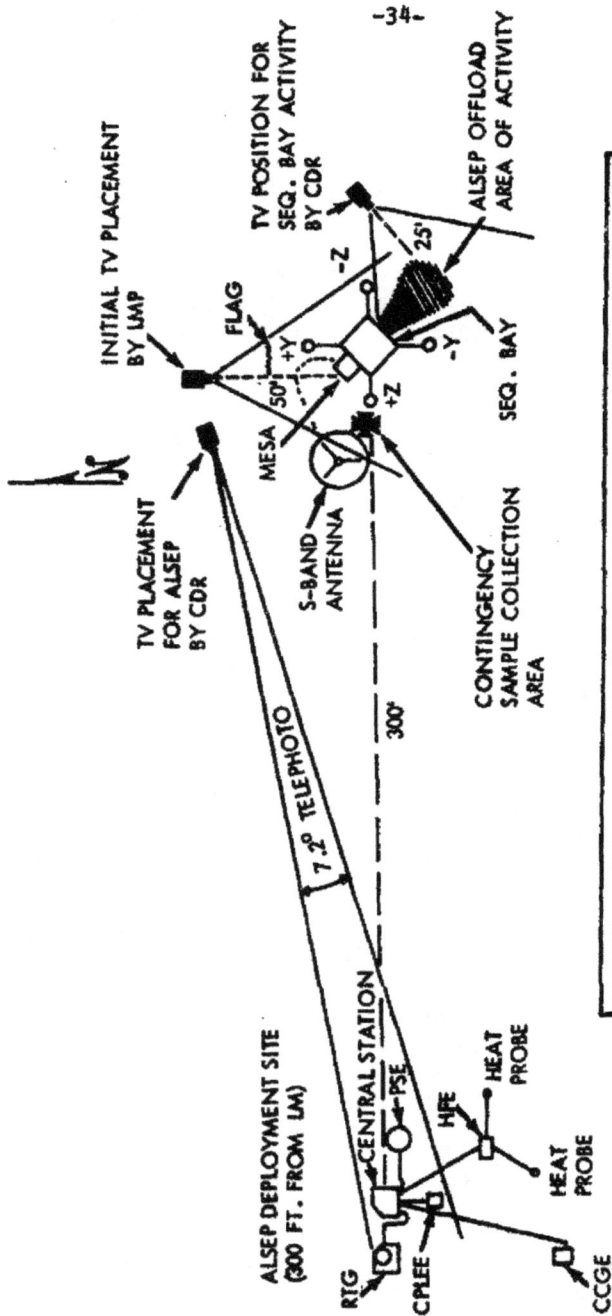

EVA-1 SURFACE ACTIVITY

LEGEND:
LMP — LM PILOT
CDR — COMMANDER
ALSEP — APOLLO LUNAR SURFACE EXPERIMENTS PACKAGE
SEQ — SCIENTIFIC EQUIPMENT
PSE — PASSIVE SEISMIC EXPERIMENT
HFE — HEAT FLOW EXPERIMENT
CPLEE — CHARGED PARTICLE LUNAR ENVIRONMENT EXPERIMENT
CCGE — COLD CATHODE GAUGE EXPERIMENT
RTG — RADIOISOTOPIC THERMOELECTRIC GENERATOR

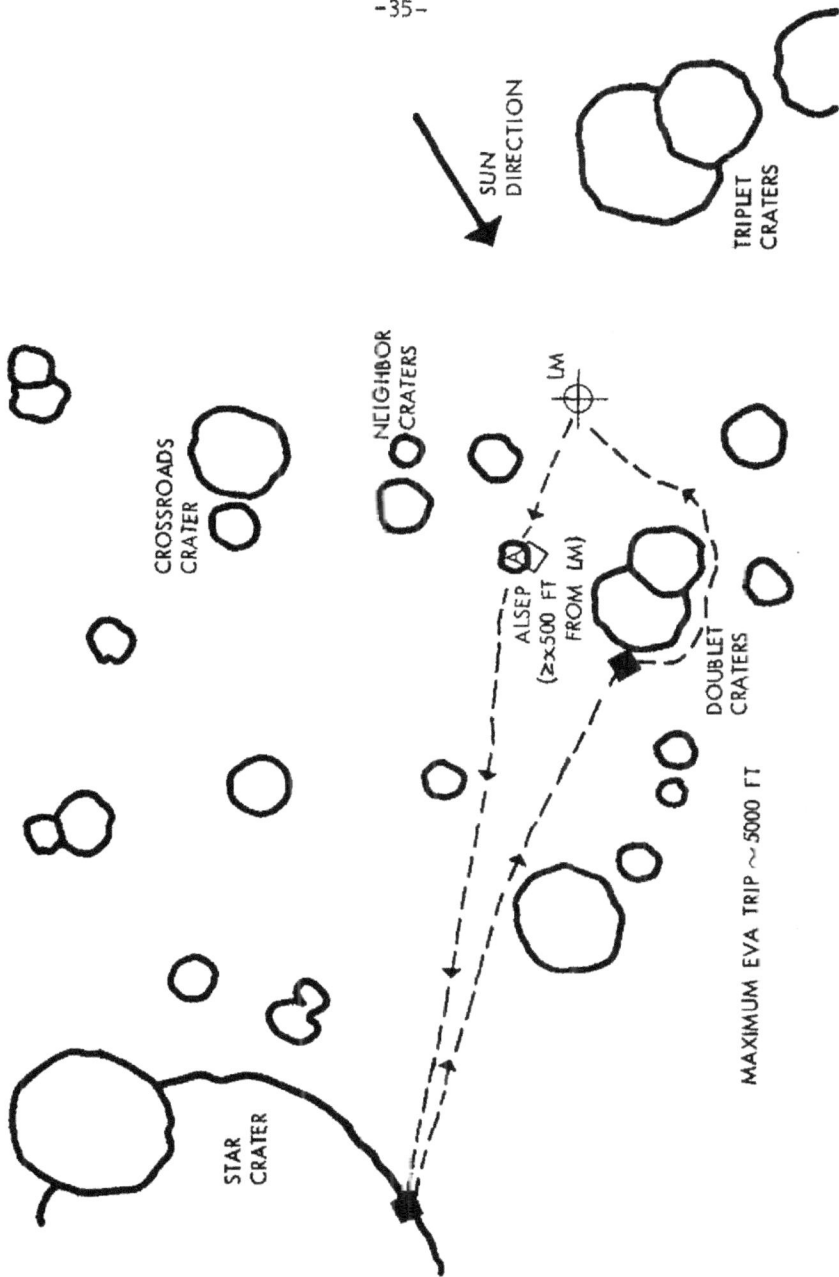

EVA-1 TRAVERSE

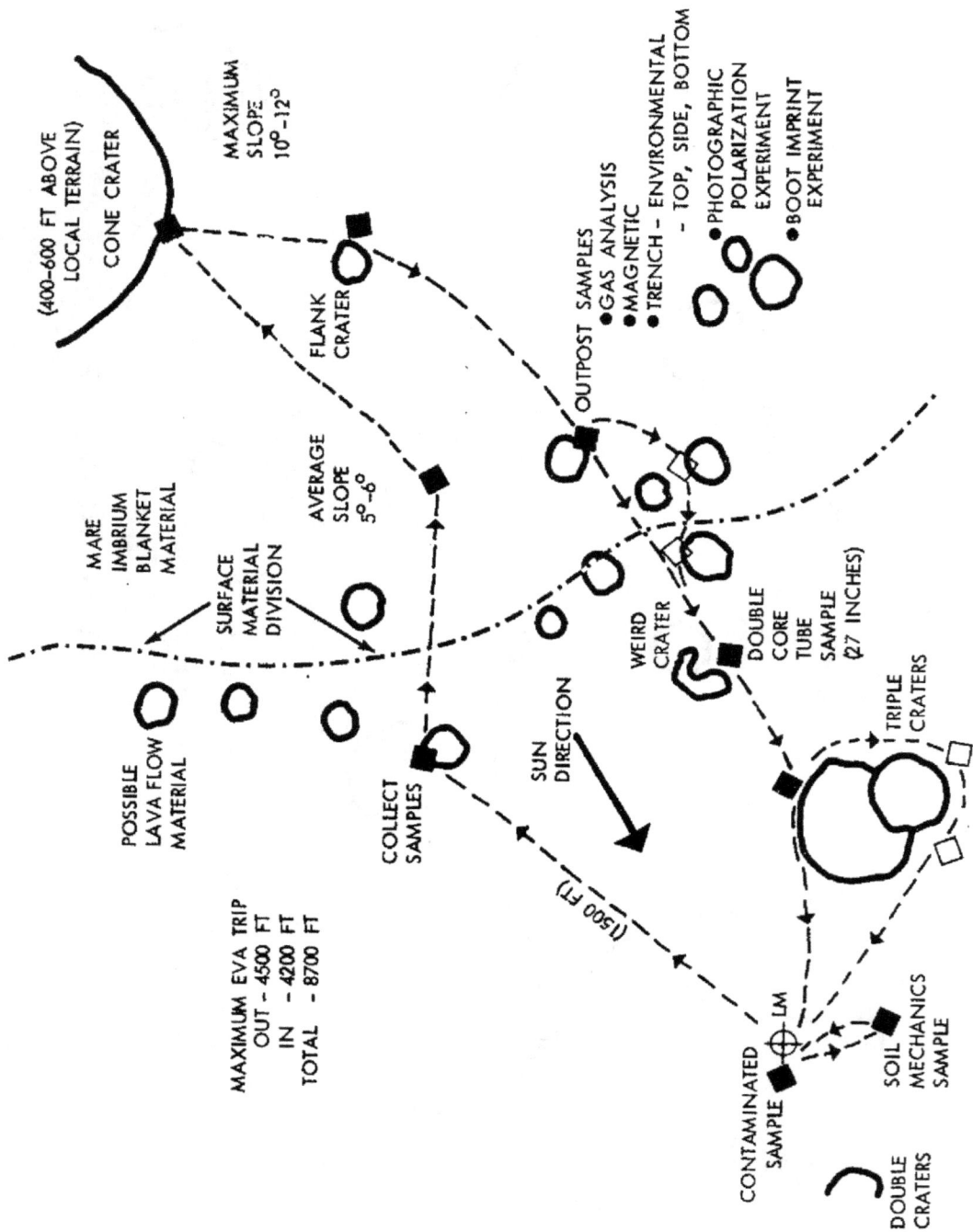

EVA-2 TRAVERSE

CONE CRATER
(400-600 FT ABOVE LOCAL TERRAIN)

MAXIMUM SLOPE 10°-12°

FLANK CRATER

OUTPOST SAMPLES
• GAS ANALYSIS
• MAGNETIC
• TRENCH - ENVIRONMENTAL
  - TOP, SIDE, BOTTOM
• PHOTOGRAPHIC POLARIZATION EXPERIMENT
• BOOT IMPRINT EXPERIMENT

MARE IMBRIUM BLANKET MATERIAL

AVERAGE SLOPE 5°-6°

SURFACE MATERIAL DIVISION

WEIRD CRATER

DOUBLE CORE TUBE SAMPLE (27 INCHES)

POSSIBLE LAVA FLOW MATERIAL

COLLECT SAMPLES

SUN DIRECTION

TRIPLE CRATERS

MAXIMUM EVA TRIP
OUT - 4500 FT
IN - 4200 FT
TOTAL - 8700 FT

(1500 FT)

LM

CONTAMINATED SAMPLE

SOIL MECHANICS SAMPLE

DOUBLE CRATERS

Red stripes around the elbows and knees of Lovell's pressure suit will permit crew recognition during EVA television transmissions and on photographs.

Ascent, Lunar Orbit Rendezvous

Following the 33-hour lunar stay the LM ascent stage will lift off the lunar surface to begin the rendezvous sequence with the orbiting CSM. Ignition of the LM ascent engine will be at 137:09 for a seven minute eight second burn attaining a total velocity of 6,044 fps. Powered ascent is in two phases: vertical ascent for terrain clearance and the orbital insertion phase. Pitchover along the desired launch azimuth begins as the vertical ascent rate reaches 50 fps about 10 seconds after liftoff at about 272 feet in altitude. Insertion into a 9x44 nm lunar orbit will take place about 166 nm west of the landing site.

Following LM insertion into lunar orbit, the LM crew will compute onboard the major maneuvers for rendezvous with the CSM which is about 267 nm ahead of and 51 miles above the LM at this point. All maneuvers in the sequences will be made with the LM RCS thrusters. The premission rendezvous sequence maneuvers, time, and velocities, which likely will differ slightly in real time, are as follows:

Concentric sequence initiate (CSI): At first LM apolune after insertion, 138:19 GET, 50 fps posigrade, following some 20 minutes of LM rendezvous radar tracking and CSM sextant/VHF ranging navigation. CSI will be targeted to place the LM in an orbit 15 nm below the CSM at the time of the later constant delta height (CDH) maneuver (139:04).

The CSI burn may also initiate corrections for any out-of-plane dispersions resulting from insertion azimuth errors. The resulting LM orbit after CSI will be 45x43.5 nm and will have a catchup rate to the CSM of about 120 feet per second.

Terminal phase initiation (TPI): This maneuver occurs at 139:46 and adds 24.7 fps along the line of sight toward the CSM when the elevation angle to the CSM reaches 26.6 degrees. The LM orbit becomes 61x44 nm and the catchup rate to the CSM decreases to a closing rate of 133 fps.

Midcourse correction maneuvers will be made if needed, followed by four braking maneuvers. Docking nominally will take place at 140:25 GET to end the three and one-half hour rendezvous sequence.

The LM ascent stage will be jettisoned at 143:04 GET and a CSM RCS 1.0 fps maneuver will provide separation.

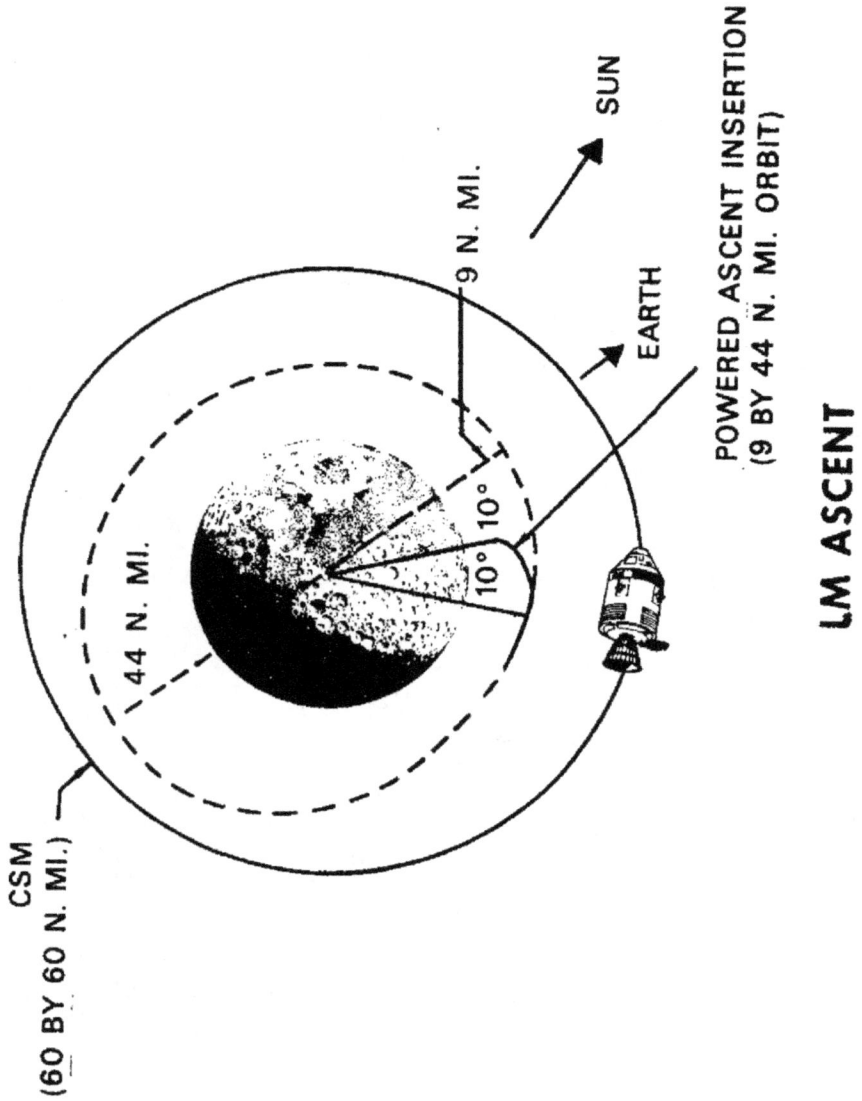

CSM
(60 BY 60 N. MI.)

44 N. MI.

10°

10°

10°

9 N. MI.

SUN

EARTH

POWERED ASCENT INSERTION
(9 BY 44 N. MI. ORBIT)

LM ASCENT

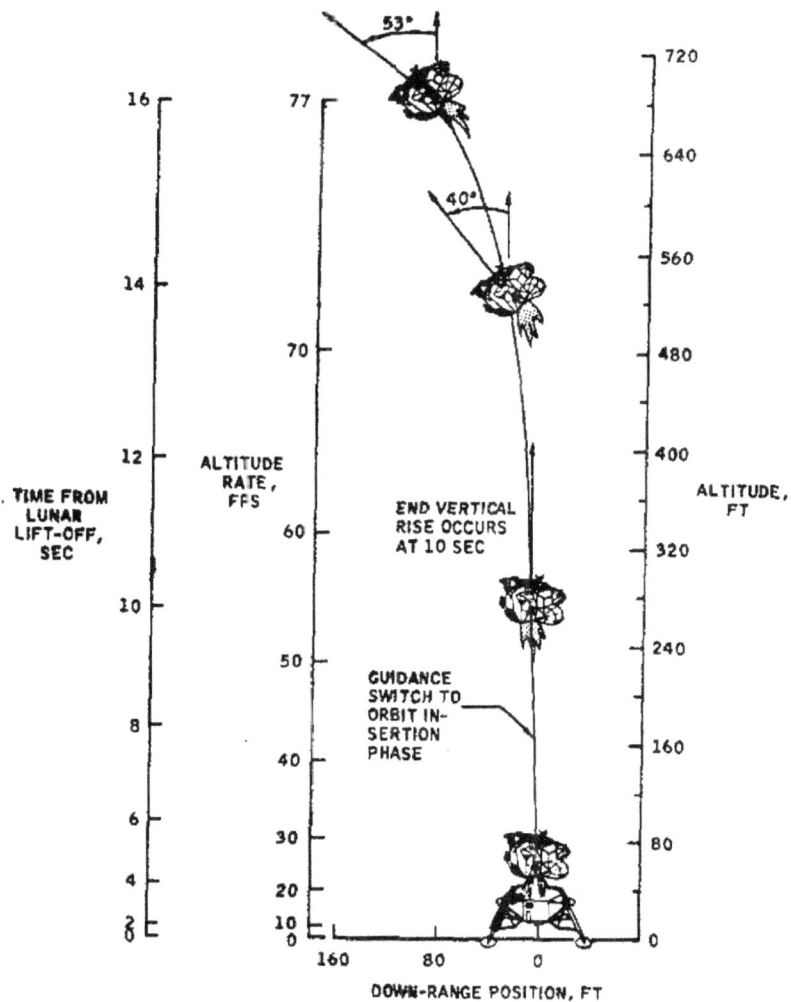

VERTICAL RISE PHASE

ORBIT INSERTION PHASE

CSM attitudes shown are for illustration purposes only

ALTITUDE (N. MI. INCREMENTS)

SURFACE RANGE, N. MI.

ALTITUDE, N. MI.

EARTH

SUN

LIFT-OFF

ORBIT INSERTION

SUMMARY

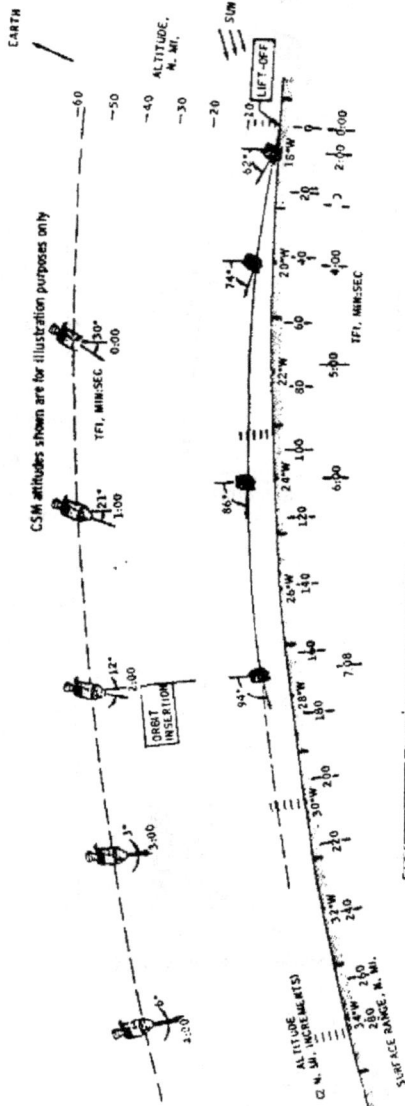

EVENT	TFI, MIN:SEC	INERTIAL VELOCITY, FPS	ALTITUDE RATE, FPS	ALTITUDE, FT	LM TO CSM			
					RANGE, N. MI.	RANGE RATE, FPS	PHASE ANGLE, DEG	LOOK ANGLE, (LOCAL VERTICAL) DEG
LIFT-OFF	0:00	15	0	0	87	3840	3.8	49.5
END OF VERTICAL RISE	0:10	56	54	271	94	4015	4.3	53.3
	2:00	1043	173	14356	168	3938	9.4	75.1
	4:00	2154	185	36588	232	2543	13.4	81.6
	6:00	4310	105	55175	265	768	15.4	87.0
ORBIT INSERTION	7:28	5533.4	34.0	59951.4	297.2	-447.3	15.6	97.3

$h_a = 53,666.5$ ft
$h_p = 43.7$ N. MI.
$\theta_a = 20.2°$
$T = 35°$
$\Delta V = 6045.2$ FPS

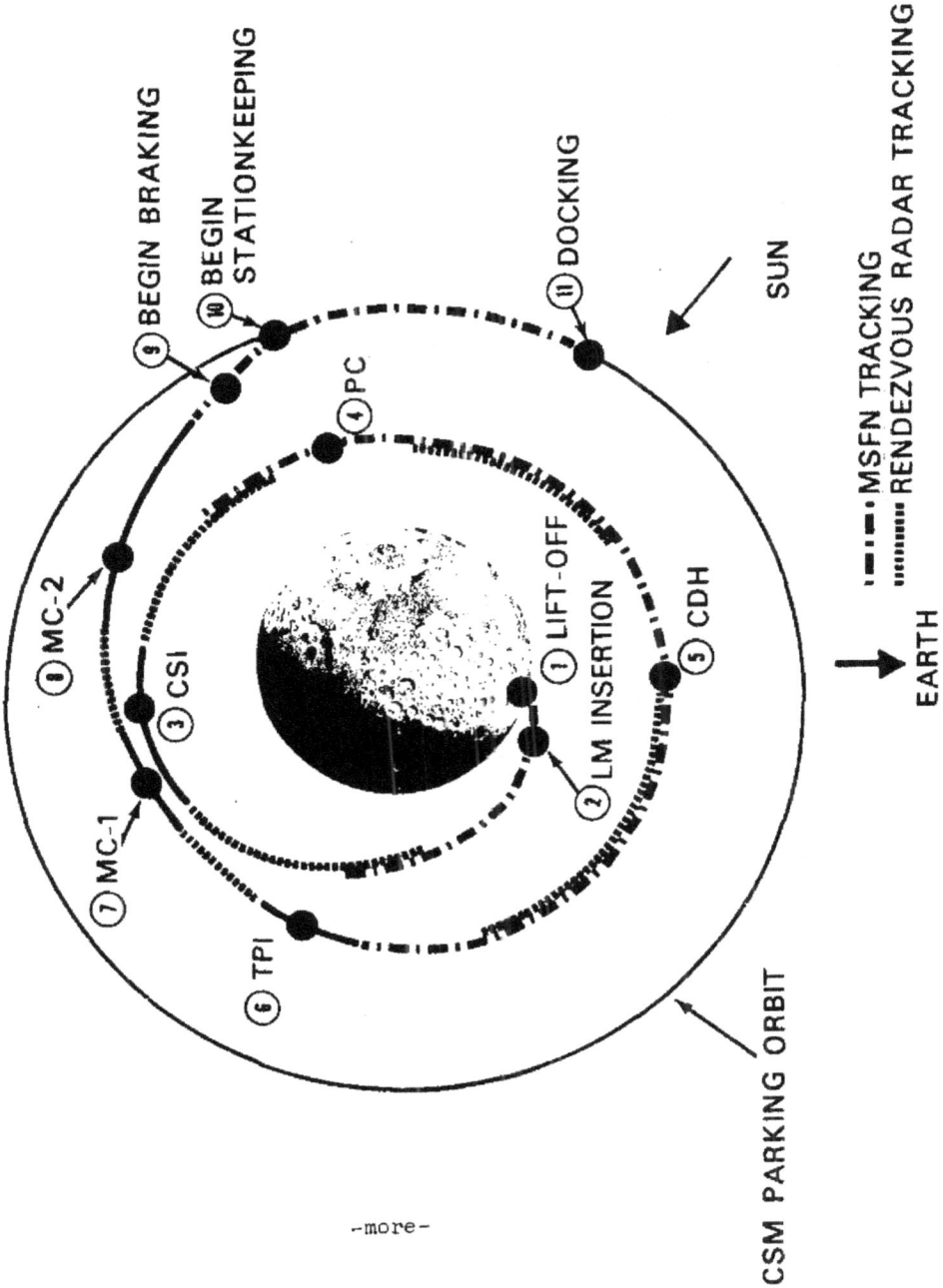

ASCENT THROUGH DOCKING

Ascent Stage Deorbit

Prior to transferring to the command module, the LM crew will set up the LM guidance system to maintain the ascent stage in an inertial attitude. At about 144:32 GET the LM RCS thrusters will ignite on ground command for 186 fps retrograde burn targeted for ascent stage impact at 145:00 about 35 miles from the landing site. The burn will have a small out-of-plane north component so that the ground track will include the original landing site. The ascent stage will impact at about 5508 fps at an angle of four degrees relative to the local horizontal. The ascent stage deorbit serves to remove debris from lunar orbit. Impacting an object with a known velocity and mass near the landing site will provide experimenters with an event for calibrating readouts from the ALSEP seismometer left behind.

A plane change maneuver at 154:13 GET will place the CSM on an orbital track passing directly over the crater Descartes and Davy Rille eight revolutions later for photographs from orbit. The maneuver will be a 825 fps/SPS burn out of plane for a plane change of 8.8 degrees, and will result in an orbit inclination of 11.4 degrees.

Transearth Injection (TEI)

The nominal transearth injection burn will be at 167:29 GET following 90 hours in lunar orbit. TEI will take place on the lunar farside, will be a 3,147 fps posigrade SPS burn of two minutes 15 seconds duration and will produce an entry velocity of 36,129 fps after a 72 hours transearth flight time.

Transearth Coast

Three entry corridor-control transearth midcourse correction burns will be made if needed: MCC-5 at TEI+15 hours, MCC-6 at entry interface (EI) -22 hours and MCC-7 at EI -3 hrs.

Entry, Landing

Apollo 13 will encounter the Earth's atmosphere (400,000 feet) at 240:50 GET at a velocity of 36,129 fps and will land approximately 1,250 nm downrange from the entry-interface point using the spacecraft's lifting characteristics to reach the landing point. Splashdown will be at 241:04 at 1.5 degrees South latitude by 157.5 degrees West longitude.

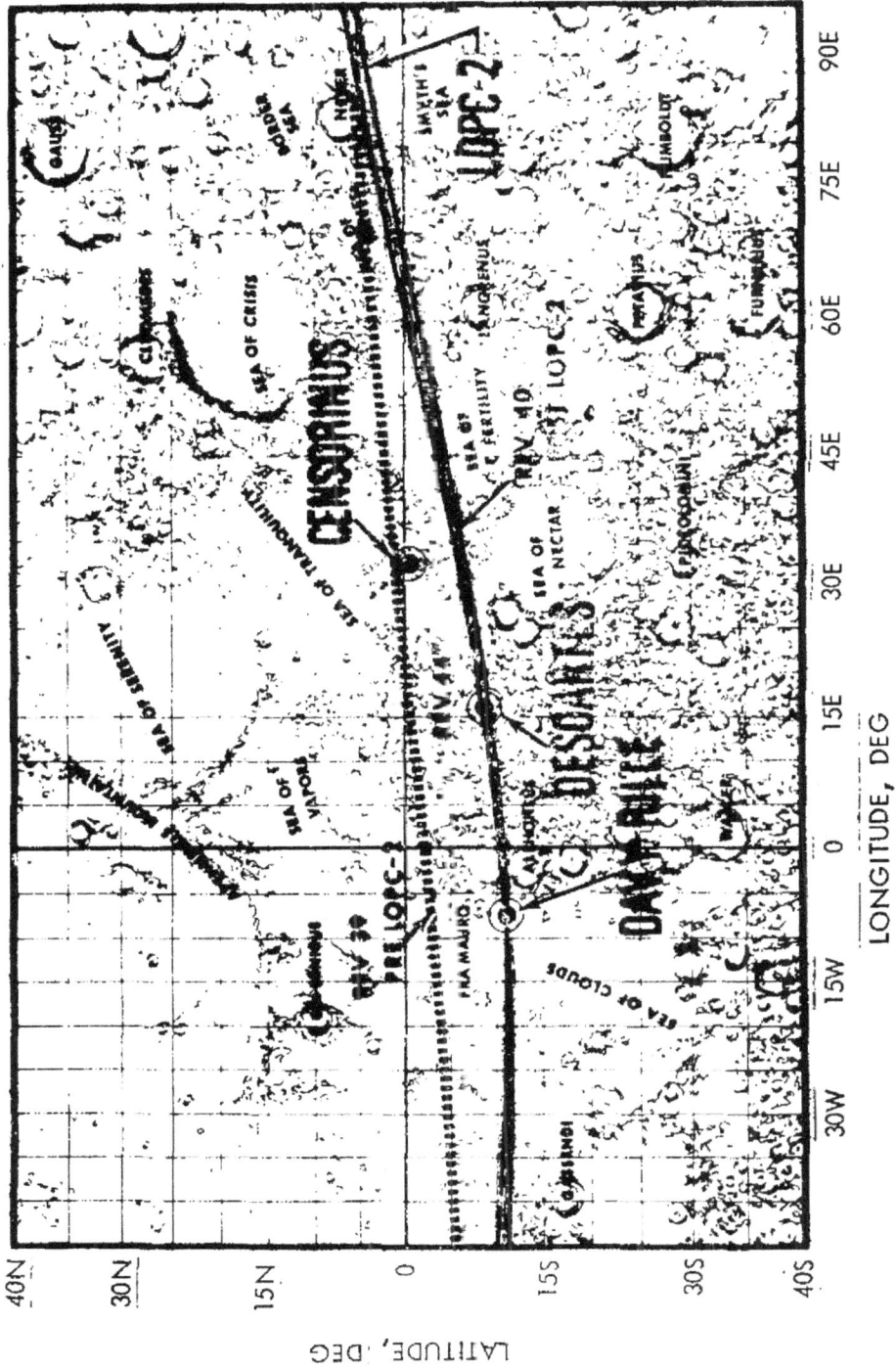

APOLLO 13 LOPC-2 TARGET LUNAR ORBIT

Recovery Operations

Launch abort landing areas extend downrange 3,400 nautical miles from Kennedy Space Center, fanwise 50 nm miles above and below the limits of the variable launch azimuth (72-96 degrees) in the Atlantic Ocean. On station in the launch abort area will be the destroyer USS New.

The landing platform-helicopter (LPH) Iwo Jima, Apollo 13 prime recovery ship, will be stationed near the Pacific Ocean end-of-mission aiming point prior entry.

Splashdown for a full-duration lunar landing mission launched on time April 11 will be at one degree 34 minutes South by 157 degrees 30 minutes West about 180 nautical miles South of Christmas Island, at 241:04 GET (3:17 p.m. EST) April 21.

In addition to the primary recovery vessel located on the mid-Pacific recovery line and the surface vessel in the launch abort area, eight HC-130 aircraft will be on standby at five staging bases around the Earth: Guam; Hawaii; Azores; Ascension Island;and Florida.

Apollo 13 recovery operations will be directed from the Recovery Operations Control Room in the Mission Control Center, supported by the Atlantic Recovery Control Center, Norfolk, Va., and the Pacific Recovery Control Center, Kunia, Hawaii.

After splashdown, the Apollo 13 crew will don clean coveralls and filter masks passed to them through the spacecraft hatch by a recovery swimmer. The crew will be carried by helicopter to the Iwo Jima where they will enter a Mobile Quarantine Facility (MQF) about 90 minutes after landing.

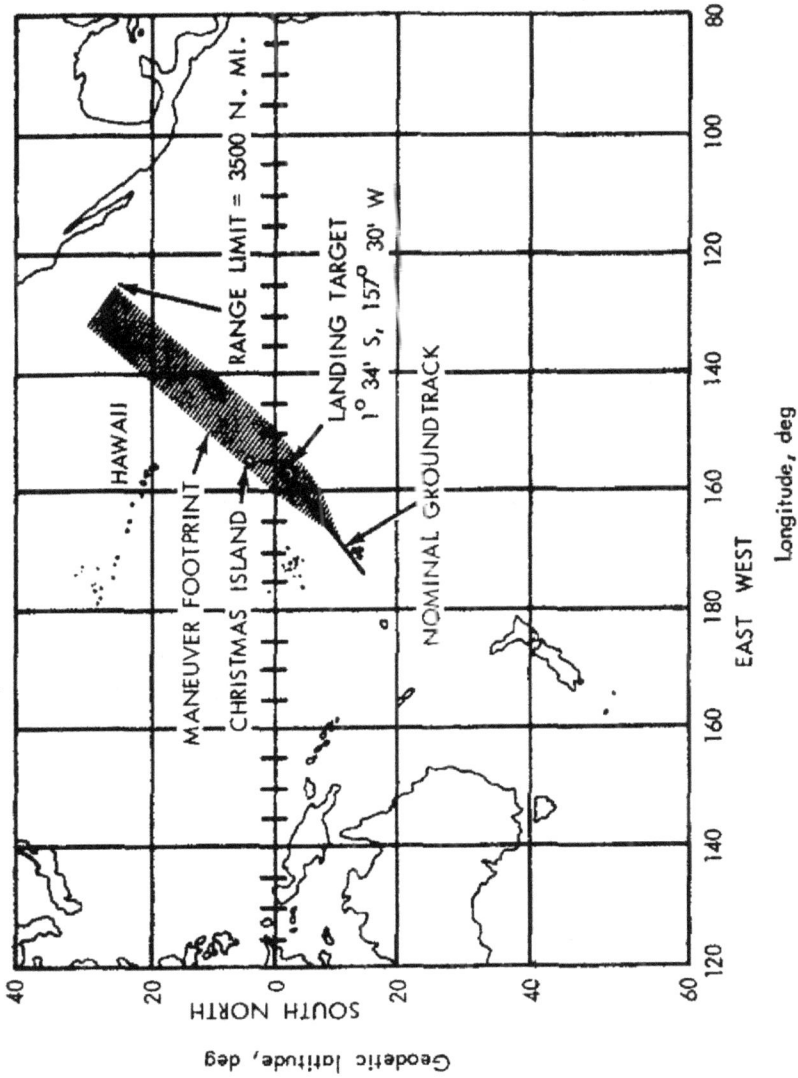

MANEUVER FOOTPRINT AND NOMINAL GROUNDTRACK

APOLLO 13 ONBOARD TELEVISION

Apollo 13 will carry two color and one black-and-white television cameras. One color camera will be used for command module cabin interiors and out-the-window Earth/Moon telecasts, and the other color camera will be stowed in the LM descent stage from where it will view the astronaut initiate egress to the lunar surface and later will be deployed on a tripod to transmit a real-time picture of the two periods of lunar surface EVA. The black-and-white camera will be carried in the LM cabin. It will only be used as a backup to the lunar surface color camera.

The two color TV cameras are essentially identical, except for additional thermal protection on the lunar surface camera. Built by Westinghouse Electric Corp., Aerospace Division, Baltimore, Md., the color cameras output a standard 525-line, 30 frame-per-second signal in color by use of a rotating color wheel system.

The color TV cameras weigh 12 pounds and are fitted with zoom lenses for wideangle or closeup fields of view. The CM camera is fitted with a three-inch monitor for framing and focusing. The lunar surface color camera has 100 feet of cable available.

The backup black-and-white lunar surface TV camera, also built by Westinghouse, is of the same type used in the first manned lunar landing in Apollo 11. It weighs 7.25 pounds and draws 6.5 watts of 24-32 volts DC power. Scan rate is 10 frames-per-second at 325 lines-per-frame. The camera body is 10.6 inches long, 6.5 inches wide and 3.4 inches deep, and is fitted with bayonet-mount wideangle and lunar day lenses.

During the two lunar surface EVA periods, Apollo 13 commander Lovell will be recognizable by red stripes around the elbows and knees of his pressure suit.

The following is a preliminary plan for TV transmissions based upon a 2:13 p.m. EST April 11 launch.

-more-

APOLLO 13 TV SCHEDULE

DAY	DATE	CST	GET	DURATION	ACTIVITY/SUBJECT	VEH	STA
Saturday	Apr. 11	2:48 pm	01:35	7 Min.	Earth	CSM	KSC **
Saturday	Apr. 11	4:28 pm	03:15	1 Hr. 8 Min.	Transposition & Docking	CSM	GDS
Sunday	Apr. 12	7:28 pm	30:15	30 Min.	Spacecraft Interior (MCC-2)	CSM	GDS
Monday	Apr. 13	11:13 pm	58:00	30 Min.	Interior & IVT to LM	CSM	GDS
Wednesday	Apr. 15	1:03 pm	95:50	15 Min.	Fra Mauro Landing Site	CSM	MAD
Thursday	Apr. 16	1:23 am	108:10	3 Hrs. 52 Min.	Lunar Surface (EVA-1)	LM	GDS/HSK
Thursday	Apr. 16	9:03 pm	127:50	6 Hrs. 35 Min.	Lunar Surface (EVA-2)	LM	GDS
Friday	Apr. 17	9:36 am	140:23	12 Min.	Docking	CSM	MAD
Saturday	Apr. 18	11:23 am	166:10	40 Min.	Lunar Surface	CSM	MAD*
Saturday	Apr. 18	1:13 pm	168:00	25 Min.	Lunar Surface (Post TEI)	CSM	MAD ●
Monday	Apr. 20	6:58 pm	221:45	15 Min.	Earth & Spacecraft Interior	CSM	GDS

●   Recorded only

**   Tentative

APOLLO 13 SCIENCE

Lunar Orbital Photography

Science experiments and photographic tasks will be conducted from the CSM during the Apollo 13 mission. During the translunar phase of the mission, photography will be taken of the Earth as well as various operational photography.

During lunar orbit, various lunar surface features including candidate landing sites Censorinus, Descartes and Davy Rille and the Apollo 11 and 12 landing sites will be photographed with the Lunar Topographic Camera. In addition, five astronomical phenomena will be photographed:

1) Photographs will be taken of the solar corona using the Moon as an occulting edge to block out the solar disk.

2) Photography will be taken of the zodiacal light which is believed to originate from reflected sunlight in the astoroid belt. Earth observation of zodiacal light is inconclusive due to atmospheric distortion.

3) Photography will be taken of lunar limb brightening, which appears as bright rim light above the horizon following lunar sunset.

4) Photographs will be taken of the Comet J.C. Bennett, 1969i which should be visible from lunar orbit during the Apollo 13 mission.

5) Photographs will be taken of the region of Gegenschein which is a faint light source covering a 20° field of view about the Earth-Sun line on the opposite side of the Earth from the Sun (anti-solar axis). One of the theories for the Gegenschein source is the existence of trapped particles of matter at the Moulton point which produce brightness due to reflected sunlight. The Moulton point is a theoretical point located 940,000 statute miles from the Earth along the anti-solar axis at which the sum of all gravitational forces is zero. From the vantage point of lunar orbit, the Moulton point region may be photographed from approximately 15° off the Earth/Sun line. These photographs should show if Gegenschein results from the Moulton point theory or from zodiacal light or a similar source.

Photographic studies will be made on Apollo 13 of the ice particle flow following a water dump and of the gaseous cloud which surrounds a manned spacecraft in a vacuum and results from liquid dumps, outgassing, etc.

In addition to the photographic studies, an experiment will be conducted with the CSM VHF communications link. During this experiment, the VHF signal will be reflected from the lunar surface and received by a 150-foot antenna on Earth. By analysis of the wavelength of the received signal, certain lunar subsurface characteristics may be discernible such as the depth of the lunar regolith layer. This experiment is called VHF Bistatic Radar.

Charged Particle Lunar Environment Experiment (CPLEE)

The scientific objective of the Charged Particle Lunar Environment Experiment is to measure the particle energies of protons and electrons that reach the lunar surface from the Sun. Increased knowledge on the energy distribution of these particles will help us understand how they perturb the Earth-Moon system. At some point electrons and protons in the magnetospheric tail of the Earth are accelerated and plunge into the terrestrial atmosphere causing the spectacular auroras and the Van Allen radiation. When the Moon is in interplanetary space the CPLEE measures proton and electrons from solar flares which results in magnetic storms in the Earth's atmosphere. Similar instruments have been flown on Javelin rockets and on satellites. The lunar surface, however, allows data to be gathered over a long period of time and from a relatively stable platform in space.

To study these phenomena, the CPLEE measures the energy of protons and electrons simultaneously from 50 electron volts to 50,000 electron volts (50Kev). The solar radiation phenomena measured are as follows:

a. Solar wind electrons and protons 50ev-5Kev.

b. Thermalized solar wind protons and electrons 50ev-10Kev.

c. Magnetospheric tail particles 50ev to 50Kev.

d. Low energy solar cosmic rays 40ev-50Kev.

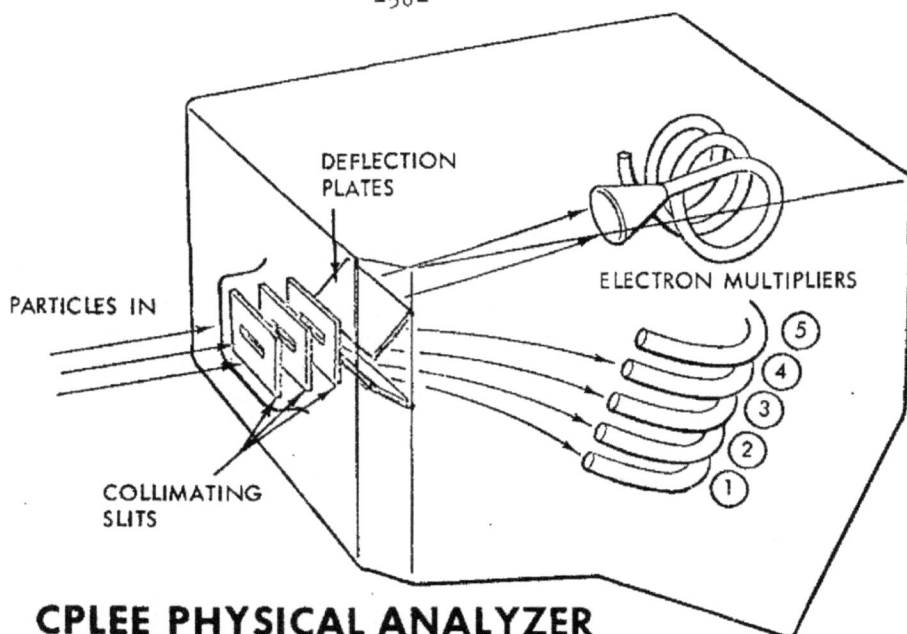

CPLEE PHYSICAL ANALYZER

PARTICLES IN

DEFLECTION PLATES

COLLIMATING SLITS

ELECTRON MULTIPLIERS

5
4
3
2
1

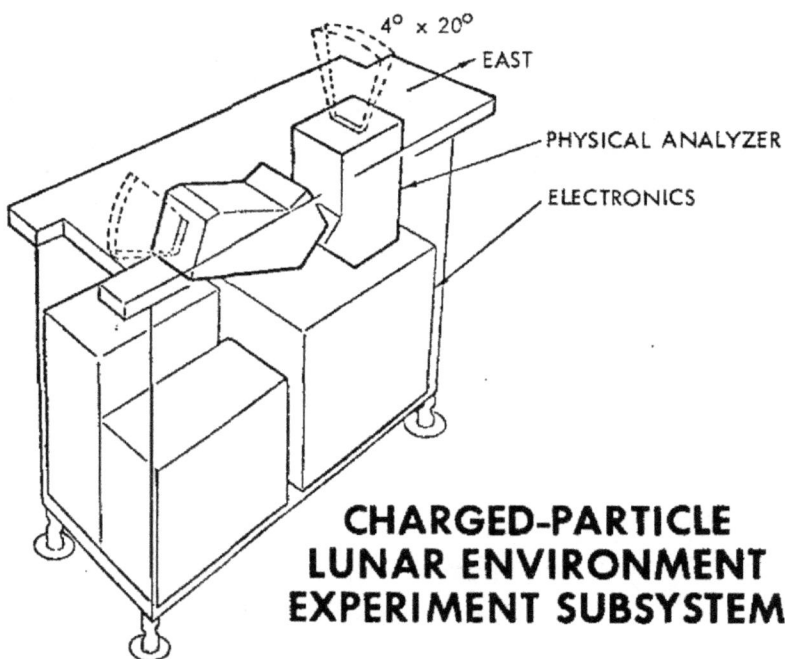

4° x 20°

EAST

PHYSICAL ANALYZER

ELECTRONICS

CHARGED-PARTICLE LUNAR ENVIRONMENT EXPERIMENT SUBSYSTEM

-more-

This experiment is distinct from the ALSEP Solar Wind Spectrometer (SWS) flown on Apollo 12 which measures direction as well as energy levels. The SWS measures electrons from 10.5ev to 1,400ev and protons from 75ev to 10,000ev.

The detector package contains two spectrometers providing data on the direction of the incoming flux.

Each spectrometer has six particle detectors: five C-shaped channeltron photon-multipliers and one funneltron, a helical shaped photon multiplier. Particles of a given charge and different energies on entering the spectrometer are subject to varying voltages and deflected toward the five channeltrons while particles of the opposite charge are deflected toward the funneltron. Thus electrons and protons are measured simultaneously in six different energy levels. The voltages are changed over six steps; +35V, +350 volts and +3500V. In this way electrons and protons are measured from 50ev to 70Kev in a period of less than 20 seconds.

The channeltron is a glass capillary tube having an inside diameter of about one millimeter and a length of 10 centimeters. The helical funneltron has an opening of 8mm. When a voltage is applied between the ends of the tube, an electric field is established down its length. Charged particles entering the tube are amplified by a factor of 10^8.

The spectrometers have two ranges of sensitivity and can measure fluxes between 10^4 and 10^{10} particles/cm^2-sec-steradian.

The charged particle lunar environment experiment (CPLEE) and data analysis are the responsibility of Dr. Brian O'Brien, University of Sydney (Australia) and Dr. David Reasoner, Rice University, with Dr. O'Brien assuming the role of Principal Investigator.

Lunar Atmosphere Detector (LAD)

Although the Moon is commonly described as a planetary body with no atmosphere, the existence of some atmosphere cannot be doubted. Two sources of this atmosphere are predicted: internal, i.e., degassing from the interior of the Moon either by constant diffusion through its surface or intermittent release from active vents; external i.e., solar wind and vaporization during meteorite impacts. Telescopic observations from polarized scattered light indicate that the atmospheric pressure could not exceed one millionth of a torr (a torr is defined as 1/760 of the standard atmosphere).

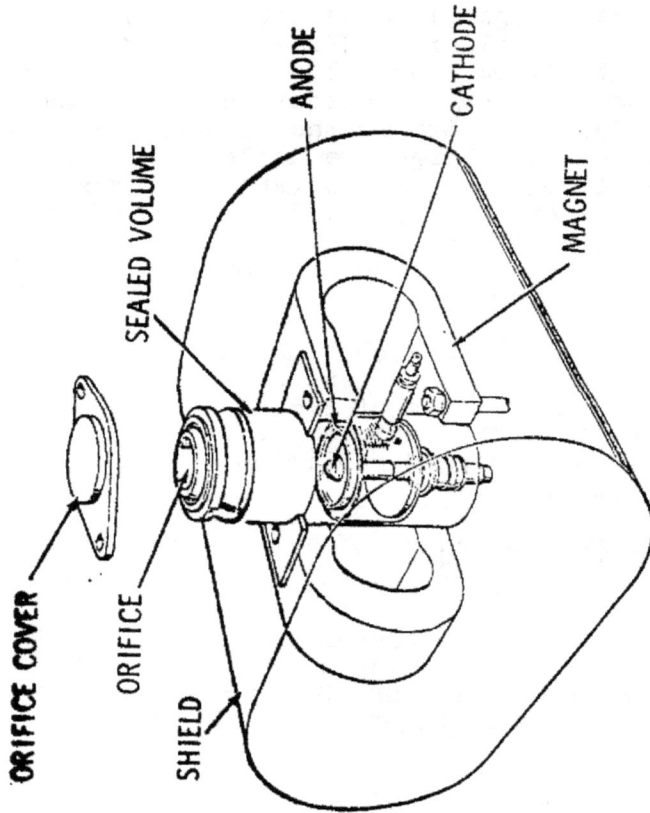

ORIFICE COVER

SEALED VOLUME

ANODE

CATHODE

MAGNET

ORIFICE

SHIELD

LUNAR ATMOSPHERIC DETECTOR
(COLD CATHODE ION GAUGE INSTRUMENT)

Measurements will be of the greatest significance if it turns out through later orbital sensors that they are of internal origin. The Earth's atmosphere and oceans have been released from the Earth's interior by degassing. The most certain source, however, is the solar wind whose ionized particles become neutralized in the lunar atmosphere and then are released as neutral gases. Neon is the predominant gas expected. Lighter gases such as hydrogen and helium escape and heavier ones statistically should be present in small quantities. Neutral particles are ionized in the lunar atmosphere, further reducing the numbers present; others will escape as the temperature rises (and concentrate near the surface when it falls).

The LAD utilizes a cold cathode ionization gauge to measure the density of neutral particles at the lunar surface and the variations in density association with lunar phase or solar activity. The ionization gauge is basically a crossed electro-magnetic field device. Electrons in the gauge are accelerated by the combined magnetic and electric fields producing a collision are collected by the cathode where they form a flow of positive ions. The positive ions current is found to be proportional to the density of the gas molecules entering the gauge. In addition, the gauge temperature is read over the range of -90° to 125°C with $\pm$ 5°C accuracy.

From the density and temperature data the pressure of the ambient lunar atmosphere can then be calculated. Chemical composition of the atmosphere however is not directly measured but the gauge has been calibrated for each gas it is expected to encounter on the lunar surface and some estimates can be made of the chemical composition. Any one of seven different dynamic ranges may be selected permitting detection of neutral particles from 10^{-6} Torr (highest pressure predicted) to 10^{-12} Torr (maximum capability of gauge). For pressure greater than 10^{-10} Torr accuracies of $\pm$ 30% will be obtained; for pressures less then 10^{-10} Torr accuracies $\pm$ 50% will be obtained. The experiment, therefore, will reduce the present uncertainty from a magnitude to a factor.

The Lunar Atmosphere Detector (LAD) and data are the responsibility of Francis Johnson, University of Texas (Dallas) and Dallas Evans, Manned Spacecraft Center, with Dr. Johnson serving as Principal Investigator.

Lunar Heat Flow Experiment (HFE)

The scientific objective of the Heat Flow experiment is to measure the steady-state heat flow from the lunar interior. Two predicted sources of heat are: 1) original heat at the time of the Moon's formation and 2) radioactivity. Scientists believe that heat could have been generated by the infalling of material and its subsequent compaction as the Moon was formed. Moreover, varying amounts of the radioactive elements uranium, thorium and potassium were found present in the Apollo 11 and 12 lunar samples which if present at depth, would supply significant amounts of heat. No simple way has been devised for relating the contribution of each of these sources to the present rate of heat loss. In addition to temperature, the experiment is capable of measuring the thermal conductivity of the lunar rock material.

The combined measurement of temperature and thermal conductivity gives the net heat flux from the lunar interior through the lunar surface. Similar measurements on Earth have contributed basic information to our understanding of volcanoes, earthquakes and mountain building processes. In conjunction with the seismic and magnetic data obtained on other lunar experiments the values derived from the heat flow measurements will help scientists to build more exact models of the Moon and thereby give us a better understanding of its origin and history.

The Heat Flow experiment consists of instrument probes, electronics and emplacement tool and the lunar surface drill. Each of two probes is connected by a cable to an electronics box which rests on the lunar surface. The electronics, which provide control, monitoring and data processing for the experiment, is connected to the ALSEP central station.

Each probe consists of two identical 20-inch (50 cm) long sections each of which contains a "gradient" sensor bridge, a "ring" sensor bridge and two heaters. Each bridge consists of four platinum resistors mounted in a thin-walled fiberglass cylindrical shell. Adjacent areas of the bridge are located in sensors at opposite ends of the 20-inch fiber-glass probe sheath. Gradient bridges consequently measure the temperature difference between two sensor locations.

-more-

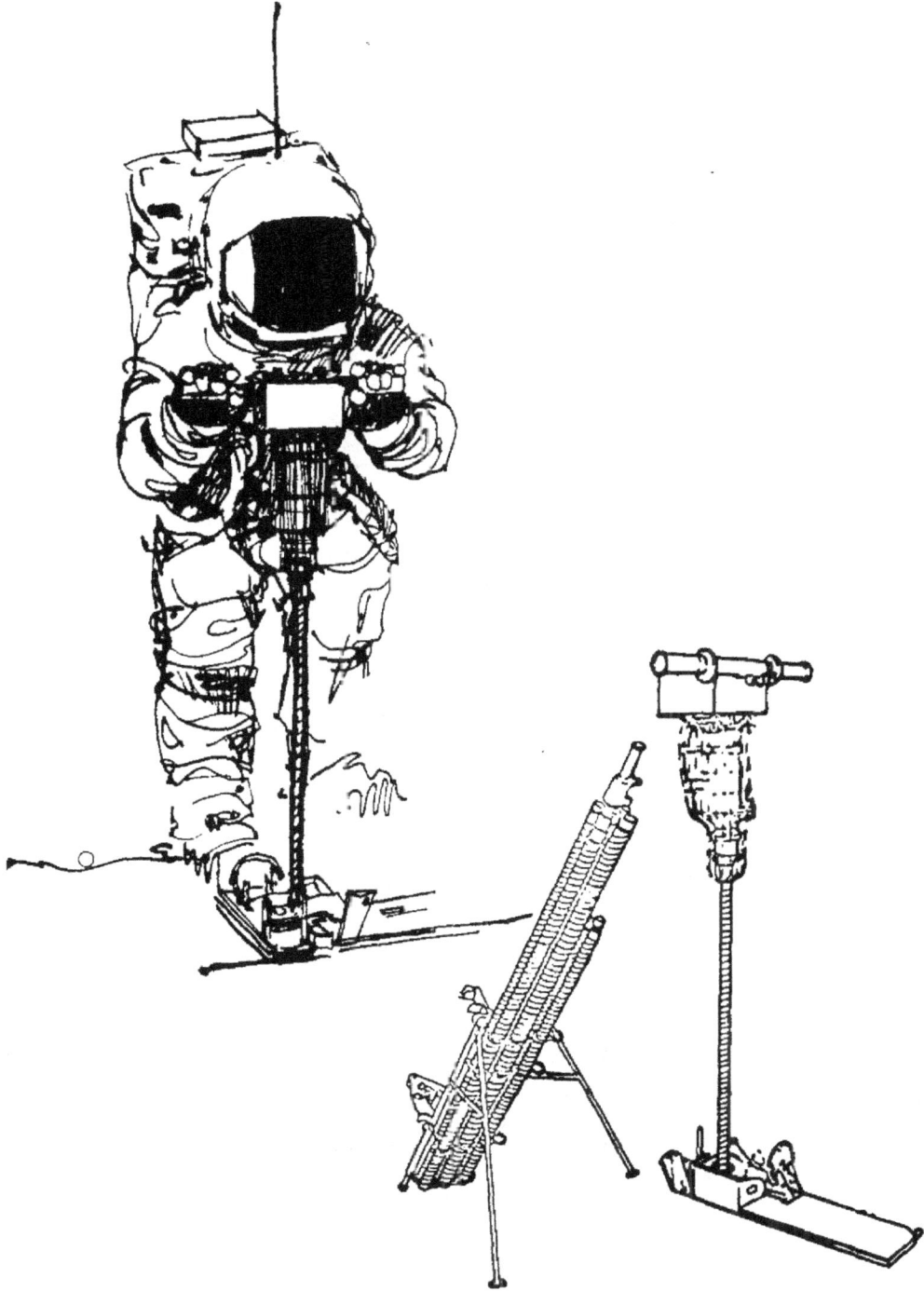

APOLLO LUNAR SURFACE DRILL

-more-

PROBE PACKAGE CABLE TRAY

ELECTRONICS PACKAGE

PROBE CARRYING PACKAGE (CONTAINS 2 PROBES & EMPLACEMENT TOOL)

SUNSHIELD

THERMAL MASK

REFLECTOR

CABLE BRACKET REMOVED DURING DEPLOYMENT

LUNAR SURFACE

TO ELECTRONICS

RADIATION SHIELD

RING SENSOR (4/PROBE)

GRADIENT SENSOR (INSIDE) 4/PROBE

HEATER COILS (OUTSIDE) PROBE STOP

RADIATION SHIELD

THERMOCOUPLES (4) 25.6, 45.3 & 65.0 IN. ABOVE PROBE

FLEXIBLE SPRING

PROBE

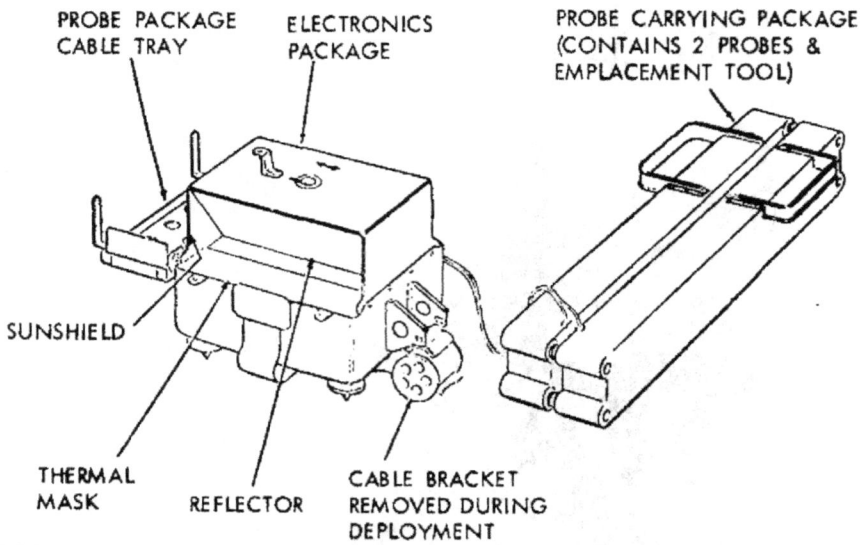

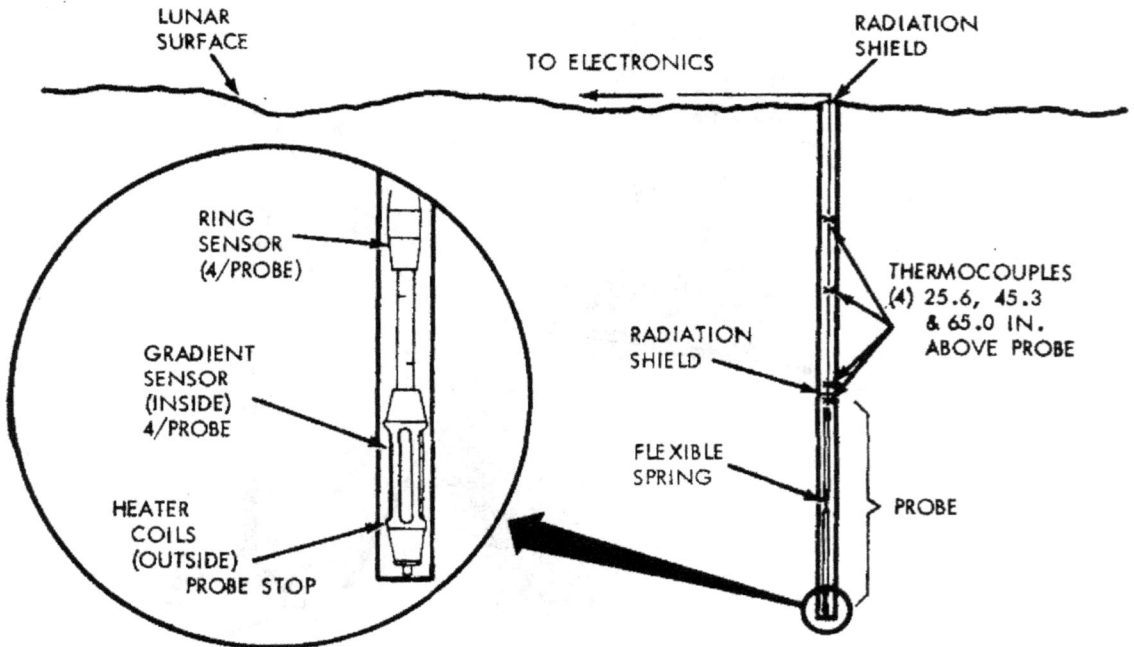

HEAT FLOW EXPERIMENT

In thermal conductivity measurements at very low values
a heater surrounding the gradient sensor is energized with
0.002 watts and the gradient sensor values monitored. The
rise in temperature of the gradient sensor is a function of
the thermal conductivity of the surrounding lunar material.
For higher range of values, the heater is energized at 0.5
watts of heat and monitored by a ring sensor. The rate of
temperature rise, monitored by the ring sensor is a function
of the thermal conductivity of the surrounding lunar material.
The ring sensor, approximately four inches from the heater, is
also a platinum resistor. A total of eight thermal conductivity measurements can be made. The thermal conductivity
mode of the experiment will be implemented about twenty days
(500 hours) after deployment. This is to allow sufficient
time for the perturbing effects of drilling and emplacing the
probe in the borehole to decay; i.e., for the probe and casings
to come to equilibrium with the lunar subsurface.

A 30-foot (10 meter) cable connects each probe to the
electronics box. In the upper six feet of the borehole the
cable contains four evenly spaced thermocouples: at the top
of the probe; at 26" (65 cm), 45" (115 cm), and 66" (165 cm).
The thermocopules will measure temperature transients pro-
pagating downward from the lunar surface. The reference junction
temperature for each thermocouple is located in the electronics
box. In fact, the feasibility of making a heat flow measurement depends to a large degree on the low thermal conductivity
of the lunar surface layer, the regolith. Measurement of lunar
surface temperature variations by Earth-based telescopes as
well as the Surveyor and Apollo missions show a remarkably
rapid rate of cooling. The wide fluctuations in temperature
of the lunar surface (from -250°F to +250°) are expected to
influence only the upper six feet and not the bottom 3 feet
of the borehole.

The astronauts will use the Apollo Lunar Surface Drill
(ALSD) to make a lined borehole in the lunar surface for the
probes. The drilling energy will be provided by a battery-
powered rotary percussive power head. The drill rod consists
of fiberglass tubular sections reinforced with boron filaments
(each about 20 inches or 50 cm long). A closed drill bit,
placed on the first drill rod, is capable of penetrating the
variety of rock including three feet of vesicular basalt
(40 per cent porosity). As lunar surface penetration pro-
gresses, additional drill rod sections will be connected to the
drill string. The drill string is left in place to serve
as a hole casing.

An emplacement tool is used by the astronaut to insert the probe to full depth. Alignment springs position the probe within the casing and assure a well-defined radiative thermal coupling between the probe and the borehole. Radiation shields on the hole prevent direct sunlight from reaching the bottom of the hole.

The astronaut will drill a third hole near the HFE and obtain cores of lunar material for subsequent analysis of thermal properties.

Heat flow experiment, design and data analysis are the responsibility of Dr. Marcus Langseth of the Lamont-Doherty Geological Observatory; Dr. Sydney Clark, Jr., Yale University, and Dr. M. G. Simmons, MIT; with Dr. Langseth assuming the role of Principal Investigator.

Passive Seismic Experiment (PSE)

The ALSEP Passive Seismic Experiment (PSE) will measure seismic activity of the Moon and obtain information on the physical properties of the lunar crust and interior. The PSE will detect surface tilt produced by tidal deformations, moonquakes and meteorite impacts.

The passive seismometer design and subsequent experiment analysis are the responsibility of Dr.Gary Latham of the Lamont-Doherty Geological Observatory.

A similar passive seismic experiment was deployed as part of the Apollo 12 ALSEP station at Surveyor crater last November and has transmitted Earthward lunar surface seismic activities since that time. The Apollo 12 and 13 seismometers differ from the seismometer left at Tranquility Base in July 1969 by the Apollo 11 crew in that they are continuously powered by a SNAP-27 radioisotope electric generator, while the Apollo 11 seismometer was powered by solar energy and could output data only during the lunar day at its location.

After Lovell and Haise ascend from the lunar surface and rendezvous with the command module in lunar orbit, the lunar module ascent stage will be jettisoned and later ground-commanded to impact on the lunar surface about 42 statute miles from the Apollo 13 landing site at Fra Mauro. Impact of an object of known mass and velocity will assist in calibrating the Apollo 13 seismometer readouts as well as providing comparative readings between the Apollo 12 and 13 seismometers forming the first two stations of a lunar surface seismic network.

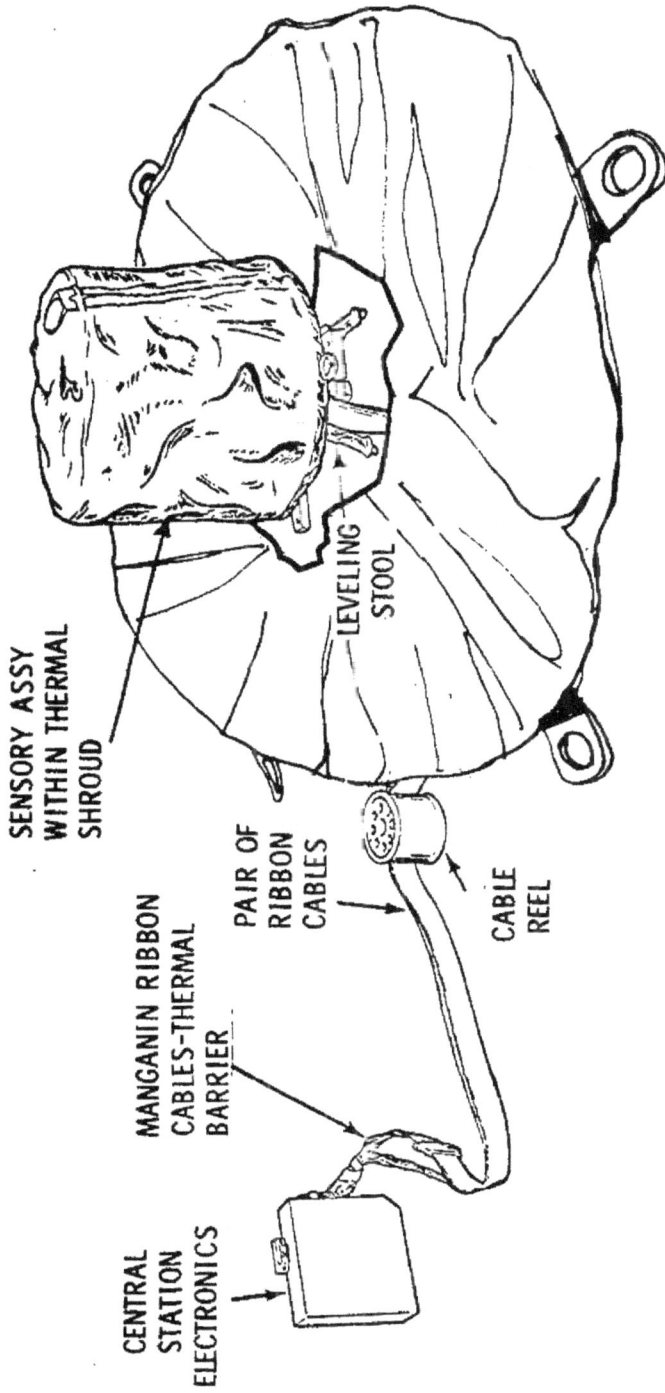

SENSORY ASSY WITHIN THERMAL SHROUD

LEVELING STOOL

MANGANIN RIBBON CABLES-THERMAL BARRIER

PAIR OF RIBBON CABLES

CABLE REEL

CENTRAL STATION ELECTRONICS

PASSIVE SEISMIC EXPERIMENT

There are three major physical components of the PSE:

* The sensor assembly consists of three, long-period seismometers with orthogonally-oriented, capacitance type seismic sensors, measuring along two horizontal axes and one vertical axis. This is mounted on a gimbal platform assembly. There is one short period seismometer which has magnet-type sensors. It is located directly on the base of the sensor assembly.

* The leveling stool allows manual leveling of the sensor assembly by the astronaut to within +5°, and final leveling to within 3 arc seconds by control motors.

* The thermal shroud covers and helps stabilize the temperature of the sensor assembly. Also, two radio-isotope heaters will protect the instrument from the extreme cold of the lunar night.

Solar Wind Composition Experiment (SWCE)

The scientific objective of the solar wind composition experiment is to determine the elemental and isotopic composition of the noble gases in the solar wind. (This is not an ALSEP experiment).

The solar wind composition detector experiment design and subsequent data analysis are the responsibility of J. Geiss and P. Eberhardt, University of Bern (Switzerland) and P. Signer, Swiss Federal Institute of Technology, with Professor Geiss assuming the responsibility of Principal Investigator.

As in Apollo 11 and 12 the SWC detector will be deployed on the Moon and brought back to Earth by the astronauts. The detector, however, will be exposed to the solar wind flux for 20 hours instead of two hours as in Apollo 11 and 18 hours 42 minutes on Apollo 12.

The solar wind composition detector consists of an aluminum foil four square feet in area and about 0.5 mils thick rimmed by Teflon for resistance to tear during deployment. A staff and yard arrangement will be used to deploy the foil and to maintain the foil approximately perpendicular to the solar wind flux. Solar wind particles will penetrate into the foil while cosmic rays will pass right through. The solar wind particles will be firmly trapped at a depth of several hundred atomic layers. After exposure on the lunar surface, the foil is reeled and returned to Earth.

SOLAR WIND EXPERIMENT

Dust Detector

The ALSEP Dust Detector is an engineering measurement designed to detect the presence of dust or debris that may impinge on the ALSEP or accumulate during its operating life.

The measurement apparatus consists of three calibrated solar cells, one pointing in east, west and vertical to face the eliptic path of the Sun. The detector is located on the central station.

Dust accumulation on the surface of the three solar cells will reduce the solar illumination detected by the cells. The temperature of each cell will be measured and compared with predicted values.

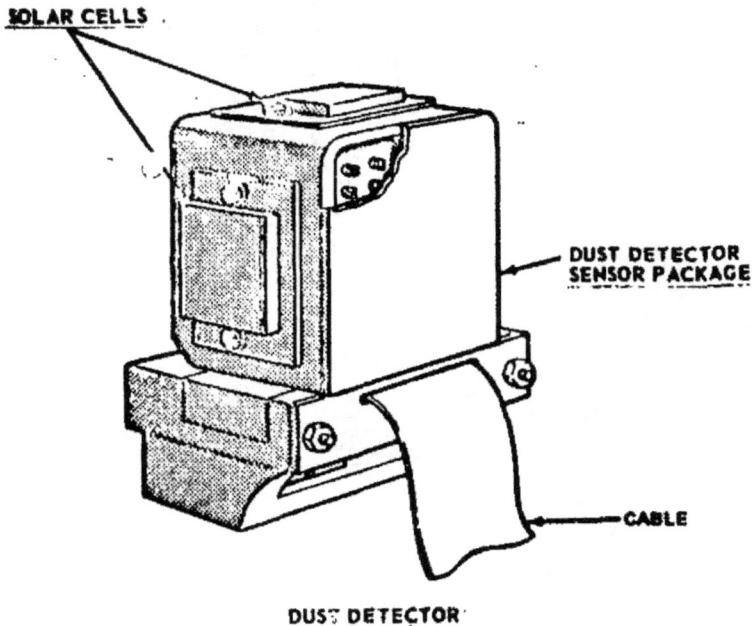

SOLAR CELLS

DUST DETECTOR
SENSOR PACKAGE

CABLE

DUST DETECTOR

Field Geology Investigations

The scientific objectives of the Apollo Field Geology Investigations are to determine the composition of the Moon and the processes which shape its surfaces. This information will help to determine the history of the Moon and its relationship to the Earth. Apollo 11 visited the Sea of Tranquility (Mare Tranquillitatis) and Apollo 12 studied the Ocean of Storms (Oceanus Procellarum). The results of these studies should help establish the nature of Mare-type areas. Apollo 13 will investigate a hilly upland area.

Geology investigation of the Moon actually began with the telescope. Systematic geology mapping began 10 years ago with a team of scientists at the U.S. Geological Survey. Ranger, Surveyor, and especially Lunar Orbiter data enormously increased the detail and accuracy of these studies. The Apollo 11 and 12 investigations represent another enormous advancement in providing new evidence on the Moon's great age, its curious chemistry, the surprisingly high density of the lunar surface material.

On Apollo 13, almost the entire second EVA will be devoted to the Field Geology Investigations and the collection of documental samples. The sample locations will be carefully photographed before and after sampling. The astronauts will carefully describe the setting from which the sample is collected. In addition to specific tasks, the astronauts will be free to photograph and sample phenomena which they judge to be unusual, significant, and interesting. The astronauts are provided with a package of detailed photo maps which they will use for planning traverses. Photographs will be taken from the LM window. Each feature or family of features will be described, relating to features on the photo maps. Areas and features where photographs should be taken and representative samples collected will be marked on the maps. The crew and their ground support personnel will consider real-time deviation from the nominal plan based upon an on-the-spot analysis of the actual situation. A trench will be dug for soil mechanics investigations.

The Earth-based geologists will be available to advise the astronauts in real-time and will work with the data returned, the photos, the samples of rock and the astronauts' observations to reconstruct here on Earth the astronauts traverse on the Moon.

Each astronaut will carry a Lunar Surface Camera (a modified 70 mm electric Hasselblad). The camera has a 60 mm lens and a Reseau plate. Lens apertures range from $f/5.6$ to $f/45$. Its focus range is from three feet to infinity. A removable polarizing filter is attached to the lens of one of the cameras and can be rotated in 45-degree increments for light polarizing studies.

A gnomon, used for metric control of near field (less than 10 feet) stereoscopic photography, will provide angular orientation relative to the local vertical. Information on the distances to objects and on the pitch, roll, and azimuth of the camera's optic axis are thereby included in each photograph. The gnomon is a weighted tube suspended vertically on a tripod supported gimbal. The tube extends one foot above the gimbal and is painted with a gray scale in bands one centimeter wide. Photogrammetric techniques will be used to produce three-dimensional models and maps of the lunar surface from the angular and distance relationship between specific objects recorded on the film.

The 16 mm Data Acquisition Camera will provide times:sequence coverage from within the LM. It can be operated in several automatic modes, ranging from one frame/second to 24 frames/second. Shutter speeds, which are independent of the frame rates, range from 1/1000 second to 1/60 second. Time exposures are also possible. While a variety of lenses is provided, the 18 mm lens will be used to record most of the geological activities in the one frame/second mode. A similar battery powered 16 mm camera will be carried in EVA.

The Lunar Surface Close-up Camera will be used to obtain very high resolution close-up stereoscopic photographs of the lunar surface to provide fine scale information on lunar soil and rock textures. Up to 100 stereo pairs can be exposed on the preloaded roll of 35 mm color film. The handle grip enables the astronaut to operate the camera from a standing position. The film drive and electronic flash are battery-operated. The camera photographs a 3"x3" area of the lunar surface.

Geological sampling equipment includes tongs, scoop, hammer, and core tubes. A 24-inch extension handle is provided for several of the tools to aid the astronaut in using them without kneeling.

Sample return containers (SRC) have been provided for return of up to 40 pounds each of lunar material for Earth-based analysis. The SRC's are identical to the ones used on the Apollo 11 and 12 missions. They are machined from aluminum forgings and are designed to maintain an internal vacuum during the outbound and return flights. The SRC's will be filled with representative samples of lunar surface material, collected and separately bagged by the astronauts on their traverse and documented by verbal descriptions and photography. Subsurface samples will be obtained by using drive tubes 16 inches long and one inch in diameter. A few grams of material will be preserved under lunar vacuum conditions in a special environmental sample container.

This container will be opened for analysis under vacuum conditions equivalent to that at the lunar surface. Special containers are provided for a magnetic sample and a gas analysis sample.

SNAP-27

SNAP-27 is one of a series of radioisotope thermoelectric generators, or atomic batteries, developed by the U.S. Atomic Energy Commission under its SNAP program. The SNAP (Systems for Nuclear Auxiliary Power) Program is directed at development of generators and reactors for use in space, on land, and in the sea.

SNAP-27 was first used in the Apollo 12 mission to provide electricity for the first Apollo Lunar Surface Experiments Package (ALSEP). A duplicate of the Apollo 12 SNAP-27 will power the Apollo 13 ALSEP.

The basic SNAP-27 unit is designed to produce at least 63 electrical watts of power. It is a cylindrical generator fueled with the radioisotope plutonium 238. It is about 18 inches high and 16 inches in diameter, including the heat radiating fins. The generator, making maximum use of the lightweight material beryllium, weighs about 28 pounds unfueled.

The fuel capsule, made of a superalloy material, is 16.5 inches long and 2.5 inches in diameter. It weighs about 15.5 pounds, of which 8.36 pounds represent fuel. The plutonium 238 fuel is fully oxidized and is chemically and biologically inert.

The rugged fuel capsule is contained within a graphite fuel cask from launch through lunar landing. The cask is designed to provide reentry heating protection and added containment for the fuel capsule in the unlikely event of an aborted mission. The cylindrical cask with hemispherical ends includes a primary graphite heat shield, a secondary beryllium thermal shield, and a fuel capsule support structure made of titanium and Inconel materials. The cask is 23 inches long and eight inches in diameter and weighs about 24.5 pounds. With the fuel capsule installed, it weighs about 40 pounds. It is mounted on the lunar module descent stage by a titanium support structure.

Once the lunar module is on the Moon, the lunar module pilot will remove the fuel capsule from the cask and insert it into the SNAP-27 generator which will have been placed on the lunar surface near the module.

The spontaneous radioactive decay of the plutonium 238 within the fuel capsule generates heat in the generator. An assembly of 442 lead telluride thermoelectric elements converts this heat -- 1480 thermal watts -- directly into electrical energy -- at least 63 watts. There are no moving parts.

Plutonium 238 is an excellent isotope for use in space nuclear generators. At the end of almost 90 years, plutonium 238 will still supply half of its original heat. In the decay process, plutonium 238 emits mainly the nuclei of helium (alpha radiation), a very mild type of radiation with a short emission range.

Before the use of the SNAP-27 system in the Apollo program was authorized, a thorough review was conducted to assure the health and safety of personnel involved in the mission and the general public. Extensive safety analyses and tests were conducted which demonstrated that the fuel would be safely contained under almost all credible accident conditions.

Contractors for SNAP-27

General Electric Co., Missile and Space Division, Philadelphia, Pa., designed, developed, and fabricated the SNAP-27 generator for the ALSEP.

The 3M Co., St. Paul, Minn., fabricated the thermoelectric elements and assembled the SNAP-27 generator.

Solar Division of International Harvester, San Diego, Calif., fabricated the generator's beryllium structure.

Hitco, Gardena, Calif., fabricated the graphite structure for the SNAP-27 Graphite LM Fuel Cask.

Sandia Corp., a subsidiary of Western Electric, operator of AEC's Sandia Laboratory, Albuquerque, N.M., provided technical direction for the SNAP-27 program.

Savannah River Laboratory, Aiken, S.C., operated by the DuPont Co. for the AEC, prepared the raw plutonium fuel.

Mound Laboratory, Miamisburg, Ohio, operated by Monsanto Research Corp., for the AEC, fabricated the raw fuel into the final fuel form and encapsulated the fuel.

SNAP-27 GENERATOR

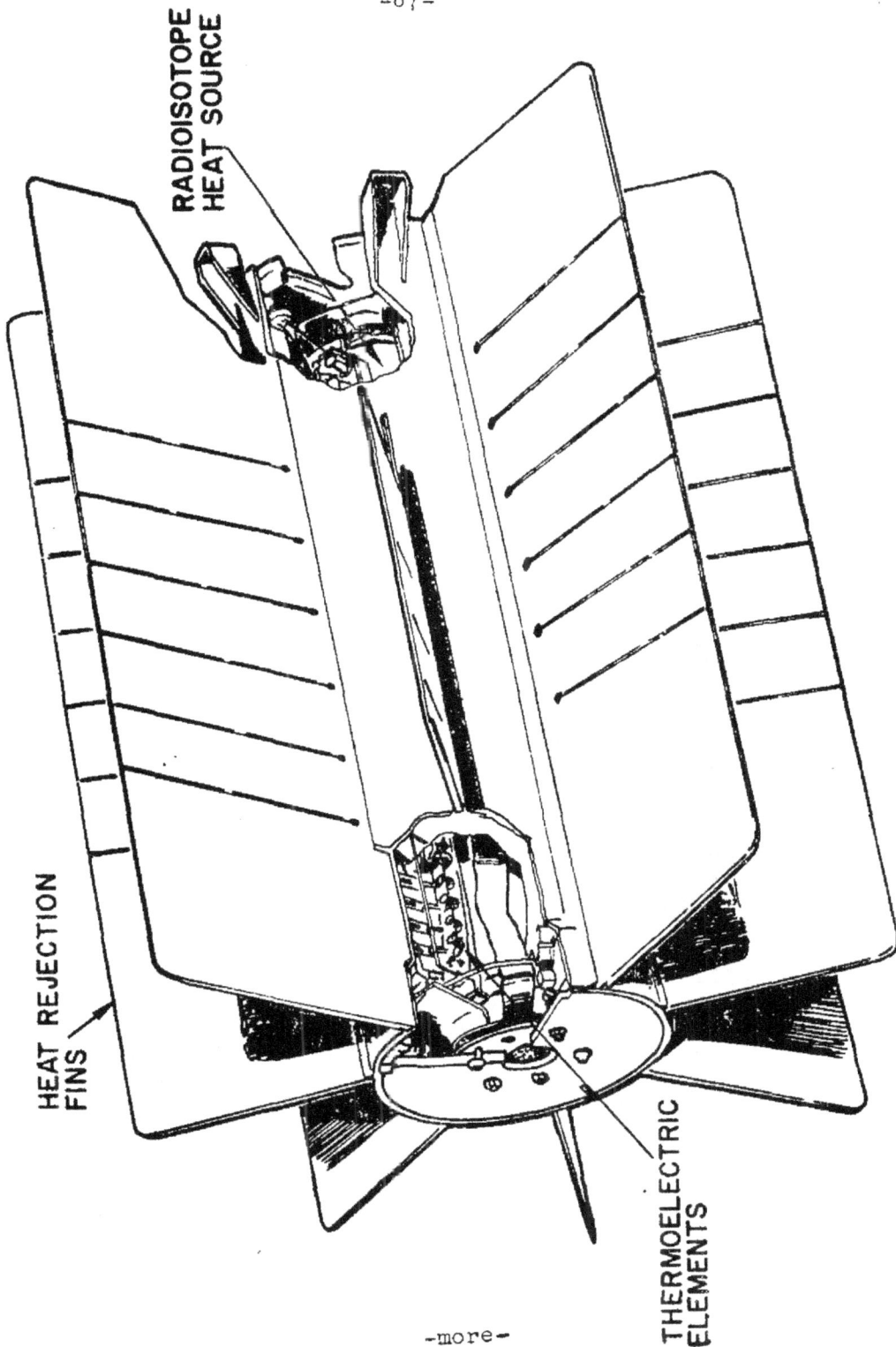

RADIOISOTOPE
HEAT SOURCE

HEAT REJECTION
FINS

THERMOELECTRIC
ELEMENTS

PHOTOGRAPHIC EQUIPMENT

Still and motion pictures will be made of most
spacecraft maneuvers and crew lunar surface activities,
and mapping photos from orbital altitude to aid in planning future landing missions. During lunar surface activities,
emphasis will be on photographic documentation of lunar
surface features and lunar material sample collection.

Camera equipment stowed in the Apollo 13 command
module consists of two 70mm Hasselblad electric cameras,
a 16mm motion picture camera, and the Hycon lunar topographic
camera (LTC).

The LTC, to be flown on Apollos 13, 14 and 15, is
stowed beneath the commander's couch. In use, the camera
mounts in the crew access hatch window.

The LTC with 18-inch focal length f/4.0 lens provides
resolution of objects as small as 15-25 feet from a 60-nm
altitude and as small as 3 to 5 feet from the 8-nm pericynthion.
Film format is 4.5-inch square frames on 100 foot long rolls,
with a frame rate variable from 4 to 75 frames a minute.
Shutter speeds are 1/50, 1/100, and 1/200 second. Spacecraft
forward motion during exposures is compensated for by a servo-
controlled rocking mount. The film is held flat in the focal
plane by a vacuum platen connected to the auxiliary dump valve.

The camera weighs 65 pounds without film, is 28 inches
long, 10.5 inches wide, and 12.25 inches high. It is a modification of an aerial reconnaissance camera.

Future lunar landing sites and targets of scientific
interest will be photographed with the lunar topographic
camera in overlapping sequence of single frame modes. A
candidate landing site northwest of the crater Censorinus
will be photographed from the 8-mile pericynthion during the
period between descent orbit insertion and CSM/LM separation.
Additional topographic photos of the Censorinus site and
sites near Davy Rille and Descartes will be made later in the
mission from the 60-nm circular orbit. The camera again will
be unstowed and mounted for 20 minutes of photography of the
lunar disc at 5 minute intervals starting at 2 hours after
transearth injection.

Cameras stowed in the lunar module are two 70mm Hasselblad data cameras fitted with 60mm Zeiss Metric lenses, a 16mm motion picture camera fitted with a 10mm lens, and a Kodak closeup stereo camera for high resolution photos on the lunar surface. The LM Hasselblads have crew chest mounts that leave both hands free.

One of the command module Hasselblad electric cameras is normally fitted with an 80mm f/2.8 Zeiss Planar lens, but bayonet mount 250mm and 500mm lenses may be substituted for special tasks.

The second Hasselblad camera is fitted with an 80mm lens and a Reseau plate which allows greater dimensional control on photographs of the lunar surface. The 500mm lens will be used only as a backup to the lunar topographic camera.

The 80mm lens has a focussing range from 3 feet to infinity and has a field of view of 38 degrees vertical and horizontal on the square-format film frame. Accessories for the command module Hasselblads include a spotmeter, intervalometer, remote control cable, and film magazines. Hasselblad shutter speeds range from time exposure and one second to one 1/500 second.

The Maurer 16mm motion picture camera in the command module has lenses of 5, 18, and 75mm available. The camera weighs 2.8 pounds with a 130-foot film magazine attached. Accessories include a right-angle mirror, a power cable, and a sestant adapter which allows the camera to use the navigation sextant optical system. The LM motion picture camera will be mounted in the right-hand window to record descent and landing and the two EVA periods and later will be taken to the surface.

The 35 mm stereo closeup camera stowed in the LM MESA shoots 24mm square stereo pairs with an image scale of one-half actual size. The camera is fixed focus and is equipped with a stand-off hood to position the camera at the proper focus distance. A long handle permits an EVA crewman to position the camera without stooping for surface object photography. Detail as small as 40 microns can be recorded. The camera allows photography of significant surface structure which would remain intact only in the lunar environment, such as fine powdery deposits, cracks or holes, and adhesion of particles. A battery-powered electronic flash provides illumination, and film capacity is a minimum of 100 stereo pairs.

LUNAR DESCRIPTION

Terrain - Mountainous and crater-pitted, the mountains rising as high as 29 thousand feet and the craters ranging from a few inches to 180 miles in diameter. The craters are thought to be formed primarily by the impact of meteorites. The surface is covered with a layer of fine-grained material resembling silt or sand, as well as small rocks and boulders.

Environment - No air, no wind, and no moisture. The temperature ranges from 243 degrees F. in the two-week lunar day to 279 degrees below zero in the two-week lunar night. Gravity is one-sixth that of Earth. Micrometeoroids pelt the Moon since there is no atmosphere to burn them up. Radiation might present a problem during periods of unusual solar activity.

Far Side - The far or hidden side of the Moon no longer is a complete mystery. It was first photographed by a Russian craft and since then has been photographed many times, particularly from NASA's Lunar Orbiter and Apollo spacecraft.

Origin - There is still no agreement among scientists on the origin of the Moon. The three theories: (1) the Moon once was part of Earth and split off into its own orbit, (2) it evolved as a separate body at the same time as Earth, and (3) it formed elsewhere in space and wandered until it was captured by Earth's gravitational field.

Physical Facts

Diameter	2,160 miles (about 1/4 that of Earth)
Circumference	6,790 miles (about 1/4 that of Earth)
Distance from Earth	238,857 miles (mean; 221,463 minimum to 252,710 maximum)
Surface temperature	+243°F (Sun at zenith) -279°F (night)
Surface gravity	1/6 that of Earth
Mass	1/100th that of Earth
Volume	1/50th that of Earth
Lunar day and night	14 Earth days each
Mean velocity in orbit	2,287 miles-per-hour
Escape velocity	1.48 miles-per-second
Month (period of rotation around Earth)	27 days, 7 hours, 43 minutes

Landing Site

The landing site selected for Apollo 13 is located
at 3° 40' 7"S, 17° 27' 3"W, about 30 miles north of the
Fra Mauro crater. The site is in a hilly, upland region.
This will be the first Apollo landing to other than a lunar
mare, the flat dark areas of the Moon once thought to be lunar
seas. This hilly region has been designated as the Fra Mauro
formation, a widespread geological unit covering large portions
of the lunar surface around Mare Imbrium (Sea of Rains). The
Fra Mauro formation is interpreted by lunar geologists to be
an ejecta blanket of material thrown out by the event which
created the circular Mare Imbrium basin.

The interpretation of the Fra Mauro formation as ejecta
from Mare Imbrium gives rise to the expectation that surface
material originated from deep within the Moon, perhaps from a
hundred miles below the Moon's surface. If the interpretation
proves correct, it will also be possible to date the Mare Imbrium
event, believed to be a major impact, perhaps the in-fall of a
smaller Moon, which was swept up in the primordial, accretionary
evolution of the Moon. Based on this theory, rocks from the
Fra Mauro formation should predate the rocks returned from
either Apollo 11 (4.6 billion years) or Apollo 12 (3.5 billion
years) and be close to the original age of the Moon.

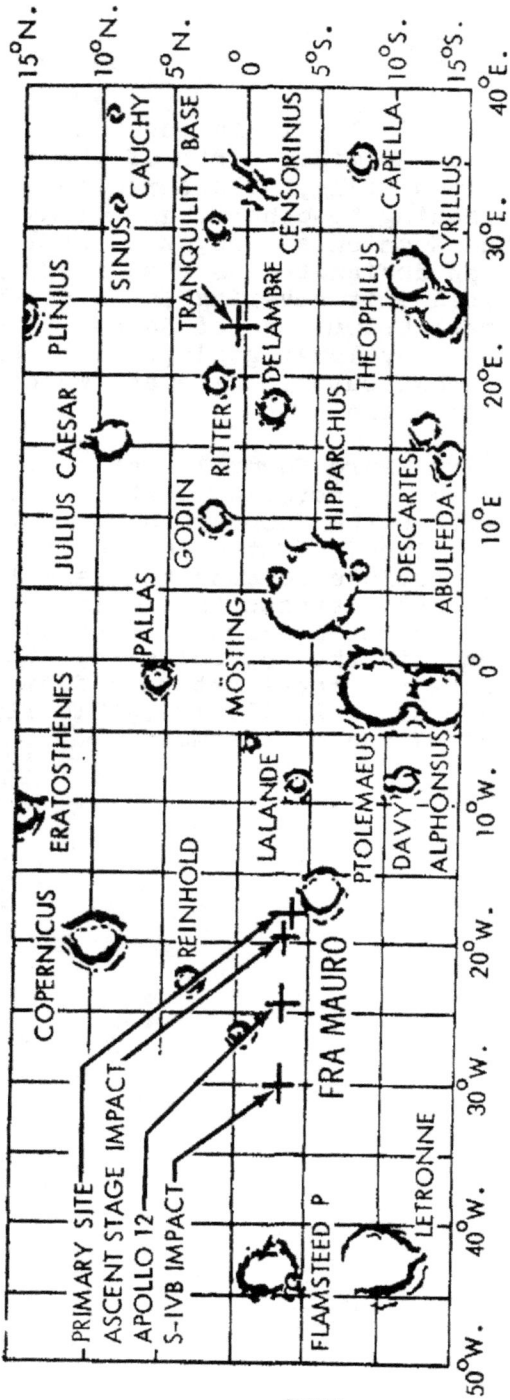

LUNAR MAP

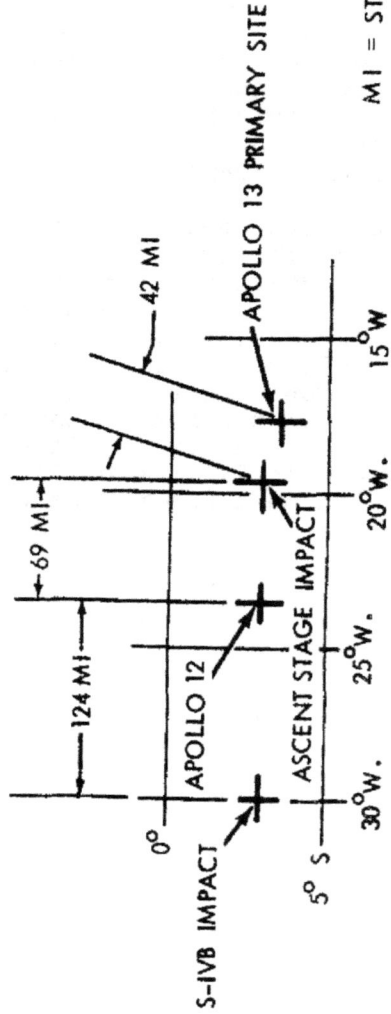

MI = STATUTE MILES

APOLLO 13 FLAGS, LUNAR MODULE PLAQUE

The United States flag to be erected on the lunar surface measures 30 by 48 inches and will be deployed on a two-piece aluminum tube eight feet long. The folding horizontal bar which keeps the flag standing out from the staff on the airless Moon has been improved over the mechanisms used on Apollo 11 and 12.

The flag, made of nylon, will be stowed in the lunar module descent stage modularized equipment stowage assembly (MESA) instead of in a thermal-protective tube on the LM front leg, as in Apollo 11 and 12.

Also carried on the mission and returned to Earth will be 25 United States and 50 individual state flags, each 4 by 6 inches.

A 7 by 9 inch stainless steel plaque, similar to those flown on Apollos 11 and 12, will be fixed to the LM front leg. The plaque has on it the words "Apollo 13" with "Aquarius" beneath, the date, and the signatures of the three crewmen.

-more-

SATURN V LAUNCH VEHICLE

The Saturn V launch vehicle (SA-508) assigned to the
Apollo 13 mission was developed at the Marshall Space Flight
Center, Huntsville, Ala. The vehicle is almost identical
to those used in the missions of Apollo 8 through 12.

First Stage

The first stage (S-IC) of the Saturn V is built by the
Boeing Company at NASA's Michoud Assembly Facility, New
Orleans, La. The stage's five F-1 engines develop a total
of about 7.6 million pounds of thrust at launch. Major components of the stage are the forward skirt, oxidizer tank,
intertank structure, fuel tank, and thrust structure. Propellant to the five engines normally flows at a rate of
29,364.5 pounds (3,400 gallons) each second. One engine is
rigidly mounted on the stage's centerline; the other four
engines are mounted on a ring at 90° angles around the center
engine. These four outer engines are gimbaled to control the
vehicle's attitude during flight.

Second Stage

The second stage (S-II) is built by the Space Division of
the North American Rockwell Corporation at Seal Beach, Calif.
Five J-2 engines develop a total of about 1.16 million pounds
of thrust during flight. Major structural components are the
forward skirt, liquid hydrogen and liquid oxygen tanks
(separated by an insulated common bulkhead), a thrust structure,
and an interstage section that connects the first and second
stages. The five engines are mounted and used in the same way
as the first stage's F-1 engines: four outer engines can be
gimbaled; the center one is rigid.

Third Stage

The third stage (S-IVB) is built by the McDonnell Douglas
Astronautics Company at Huntington Beach, Calif. Major components are the aft interstage and skirt, thrust structure,
two propellant tanks with a common bulkhead, a forward skirt,
and a single J-2 engine. The gimbaled engine has a maximum
thrust of 230,000 pounds, and can be shut off and restarted.

SPACECRAFT 82 FT.

SATURN V LAUNCH VEHICLE -281 FT.

CM

SM

LM

INSTRUMENT
UNIT

THIRD STAGE
(S-IVB)

SECOND STAGE
(S-II)

FIRST STAGE
(S-IC)

FIRST STAGE (S-IC)

Diameter ----- 33 feet
Height ------ 138 feet
Weight ------ 5,030,141 lbs. fueled
         288,000 lbs. dry
Engines ----- Five F-1
Propellants -- Liquid oxygen (3,306,494 lbs.;
         348,343 gals.) RP-1 (kerosene)
         (1,435,647 lbs.; 215,330 gals.)
Thrust ------ 7,584,593 lbs. at liftoff

SECOND STAGE (S-II)

Diameter ----- 33 feet
Height ------ 81.5 feet
Weight ----- 1,073,944 lbs. fueled
         78,050 lbs. dry
Engines ---- Five J-2
Propellants -- Liquid oxygen (836,120 lbs.;
         88,215 gals.) liquid hydrogen
         (159,774 lbs.; 272,340 gals.)
Thrust ------ 924,207 to 1,161,315 lbs.
Interstages -- 11,465

THIRD STAGE (S-IVB)

Diameter---- 21.7 feet
Height ------ 58.3 feet
Weight ------ 259,896 lbs. fueled
         24,864 lbs. dry
Engine ------ One J-2
Propellants -- Liquid oxygen (191,532 lbs.;
         20,228 gals.) Liquid hydrogen
         (43,500 lbs., 64,145 gals.)
Thrust ------ 199,790 lbs.
Interstage --- 8,100 lbs.

INSTRUMENT UNIT

Diameter-----21.7 feet
Height ------ 3 feet
Weight -------4,482 lbs.

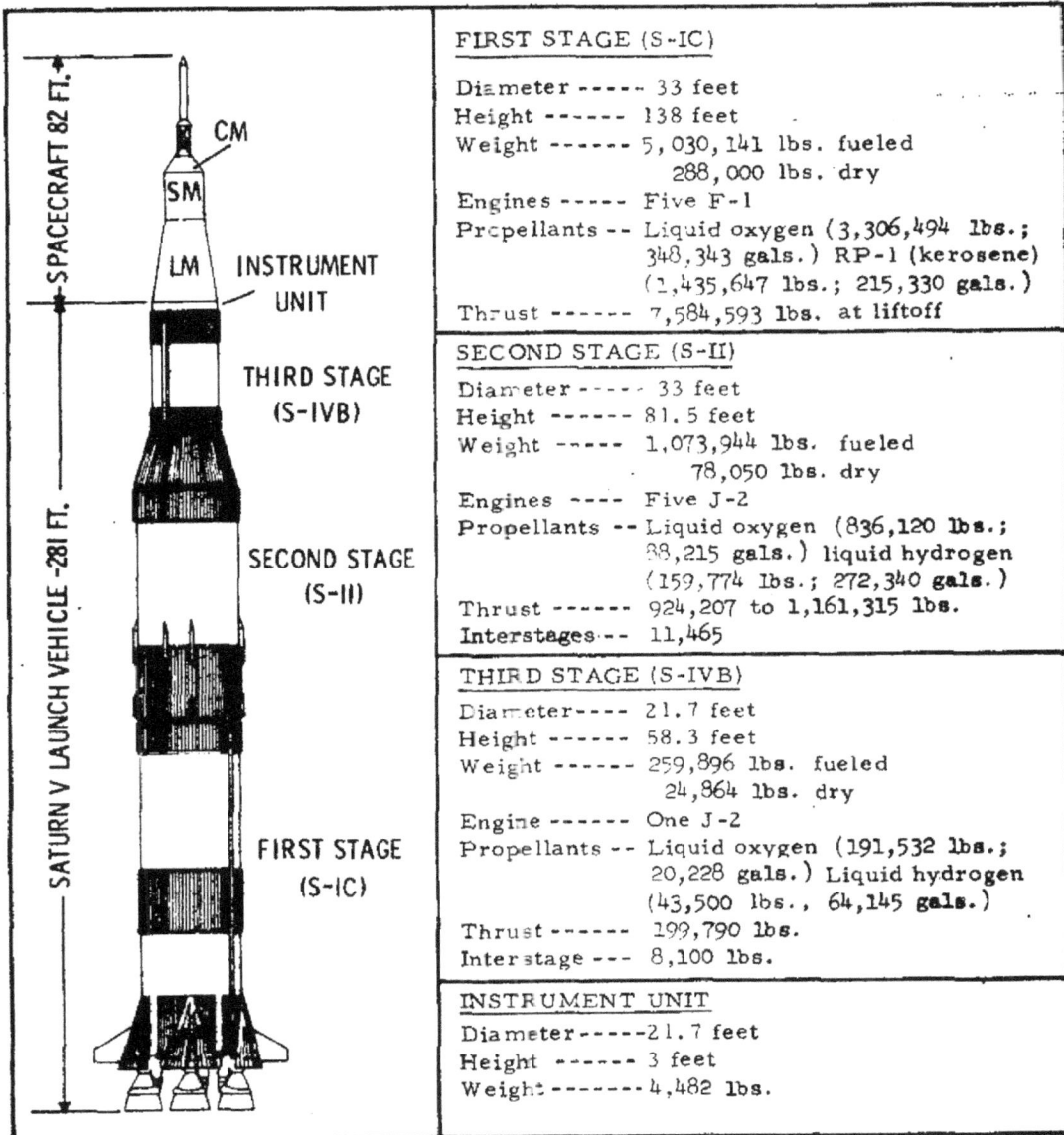

NOTE: Weights and measures given above are for the nominal vehicle configuration for Apollo 12. The figures may vary slightly due to changes before launch to meet changing conditions. Weights of dry stages and propellants do not equal total weight because frost and miscellaneous smaller items are not included in chart.

-more-

Instrument Unit

The instrument unit (IU), built by the International
Business Machines Corp., at Huntsville, Ala., contains
navigation, guidance and control equipment to steer the
launch vehicle into its Earth orbit and into translunar
trajectory. The six major systems are structural, thermal
control, guidance and control, measuring and telemetry,
radio frequency, and electric.

The instrument unit provides a path-adaptive guidance
scheme wherein a programmed trajectory is used during first
stage boost with guidance beginning during second stage burn.
This scheme prevents movements that could cause the vehicle
to break up while attempting to compensate for winds or jet
streams in the atmosphere.

The instrument unit's inertial platform (heart of the
navigation, guidance and control system) provides space-fixed
reference coordinates and measures acceleration along three
mutually perpendicular axes of a coordinate system. If the
platform fails during boost, systems in the Apollo spacecraft
are programmed to provide guidance for the launch vehicle.
After second stage ignition, the spacecraft commander could
manually steer the vehicle in the event of loss of the launch
vehicle inertial platform.

Propulsion

The Saturn V has 37 propulsive units, with thrust ratings
ranging from 70 pounds to more than 1.5 million pounds. The
large main engines burn liquid propellants; the smaller units
use solid or hypergolic propellants.

The five F-1 engines on the first stage burn a combination
of RP-1 (kerosene) as fuel and liquid oxygen as oxidizer. Each
engine develops approximately 1,516,918 pounds of thrust at
liftoff, building to about 1,799,022 pounds before cutoff.
The five-engine cluster gives the first stage a thrust range
of from 7,584,593 pounds at liftoff to 8,995,108 pounds just
before center engine cutoff. The F-1 engine weighs almost
10 tons, is more than 18 feet long and has a nozzle exit
diameter of nearly 14 feet. The engine consumes almost three
tons of propellant every second.

The first stage also has eight solid-fuel retrorockets
that fire to separate the first and second stages. Each retro-
rocket produces a thrust of 87,900 pounds for 0.6 seconds.

The second and third stages are powered by J-2 engines
that burn liquid hydrogen (fuel) and liquid oxygen (oxidizer).
J-2 engine thrust varies from 184,841 to 232,263 pounds
during flight. The 3,500-pound J-2 engine is considered
more efficient than the F-1 engine because the J-2 burns high-
energy liquid hydrogen. F-1 and J-2 engines are built by
the Rocketdyne Division of the North American Rockwell Corp.

The second stage also has four 21,000-pound-thrust solid
fuel ullage rockets that settle liquid propellant in the
bottom of the main tanks and help attain a "clean" separation
from the first stage. Four retrorockets, located in the S-IVB's
aft interstage (which never separates from the S-II), separate
the S-II from the S-IVB. There are two jettisonable ullage
rockets for propellant settling before engine ignition. Eight
smaller engines in the two auxiliary propulsion system modules
on the S-IVB stage provide three-axis attitude control.

COMMAND AND SERVICE MODULE STRUCTURE, SYSTEMS

The Apollo spacecraft for the Apollo 13 mission is comprised of Command Module 109, Service Module 109, Lunar Module 7, a spacecraft-lunar module adapter (SLA) and a launch escape system. The SLA houses the lunar module and serves as a mating structure between the Saturn V instrument unit and the SM.

Launch Escape System (LES) -- Would propel command module to safety in an aborted launch. It has three solid-propellant rocket motors: a 147,000 pound-thrust launch escape system motor, a 2,400-pound-thrust pitch control motor, and a 31,500 pound-thrust tower jettison motor. Two canard vanes deploy to turn the command module aerodynamically to an attitude with the heat-shield forward. The system is 33 feet tall and 4 feet, in diameter at the base, and weighs 8,945 pounds.

Command Module (CM) Structure -- The command module is a pressure vessel encased in heat shields, cone-shaped, weighing 12,365 pounds at launch.

The command module consists of a forward compartment which contains two reaction control engines and components of the Earth landing system; the crew compartment or inner pressure vessel containing crew accomodations, controls and displays, and many of the spacecraft systems; and the aft compartment housing ten reaction control engines, propellant tankage, helium tanks, water tanks, and the CSM umbilical cable. The crew compartment contains 210 cubic feet of habitable volume.

Heat-shields around the three compartments are made of brazed stainless steel honeycomb with an outer layer of phenolic epoxy resin as an ablative material.

CSM and LM are equipped with the probe-and-drogue docking hardware. The probe assembly is a powered folding coupling and impact attentuating device mounted in the CM tunnel that mates with a conical drogue mounted in the LM docking tunnel. After the 12 automatic docking latches are checked following a docking maneuver, both the probe and drogue are removed to allow crew transfer between the CSM and LM.

Service Module (SM) Structure -- At launch, the service module for the Apollo 13 mission will weigh 51,105 pounds. Aluminum honeycomb panels one inch thick form the outer skin, and milled aluminum radial beams separate the interior into six sections around a central cylinder containing two helium spheres, four sections containing service propulsion system fuel-oxidizer tankage, another containing fuel cells, cryogenic oxygen and hydrogen, and one sector essentially empty.

COMMAND MODULE

SERVICE MODULE

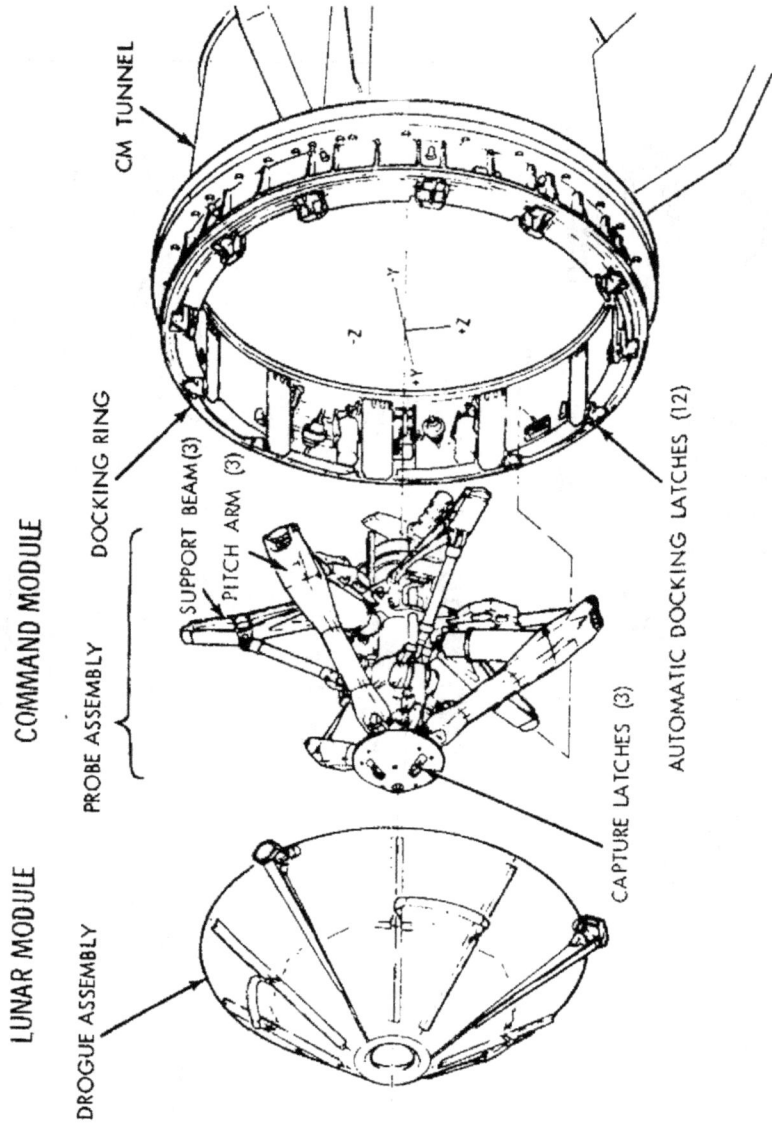

APOLLO DOCKING MECHANISMS

LUNAR MODULE   COMMAND MODULE

DROGUE ASSEMBLY   PROBE ASSEMBLY   DOCKING RING

CM TUNNEL

SUPPORT BEAM (3)

PITCH ARM (3)

CAPTURE LATCHES (3)

AUTOMATIC DOCKING LATCHES (12)

Spacecraft-LM Adapter (SLA) Structure -- The spacecraft LM adapter is a truncated cone 28 feet long tapering from 260 inches diameter at the base to 154 inches at the forward end at the service module mating line. The SLA weighs 4,000 pounds and houses the LM during launch and Earth orbital flight.

CSM Systems

Guidance, Navigation and Control System (GNCS) -- Measures and controls spacecraft position, attitude, and velocity, calculates trajectory, controls spacecraft propulsion system thrust vector, and displays abort data. The guidance system consists of three subsystems: Inertial, made up of an inertial measurement unit and associated power and data components; computer which processes information to or from other components; and optics consisting of scanning telescope and sextant for celestial and/or landmark sighting for spacecraft navigation. VHF ranging device serves as a backup to the LM rendezvous radar.

Stabilization and Control Systems (SCS) -- Controls spacecraft rotation, translation, and thrust vector and provides displays for crew-initiated maneuvers; backs up the guidance system for control functions. It has three subsystems; attitude reference, attitude control, and thrust vector control.

Service Propulsion System (SPS) -- Provides thrust for large spacecraft velocity changes through a gimbal-mounted 20,500-pound-thrust hypergolic engine using a nitrogen tetroxide oxidizer and a 50-50 mixture of unsymmetrical dimethyl hydrazine and hydrazine fuel. This system is in the service module. The system responds to automatic firing commands from the guidance and navigation system or to manual commands from the crew. The engine thrust level is not throttleable. The stabilization and control system gimbals the engine to direct the thrust vector through the spacecraft center of gravity.

Telecommunications System -- Provides voice, television, telemetry, and command data and tracking and ranging between the spacecraft and Earth, between the command module and the lunar module and between the spacecraft and astronauts during EVA. It also provides intercommunications between astronauts.

The high-gain steerable S-Band antenna consists of four, 31-inch-diameter parabolic dishes mounted on a folding boom at the aft end of the service module. Signals from the ground stations can be tracked either automatically or manually with the antenna's gimballing system. Normal S-Band voice and uplink/downlink communications will be handled by the omni and high-gain antennas.

Sequential System -- Interfaces with other spacecraft systems and subsystems to initiate time critical functions during launch, docking maneuvers, sub-orbital aborts, and entry portions of a mission. The system also controls routine spacecraft sequencing such as service module separation and deployment of the Earth landing system.

Emergency Detection System (EDS) -- Detects and displays to the crew launch vehicle emergency conditions, such as excessive pitch or roll rates or two engines out, and automatically or manually shuts down the booster and activates the launch escape system; functions until the spacecraft is in orbit.

Earth Landing System (ELS) -- Includes the drogue and main parachute system as well as post-landing recovery aids. In a normal entry descent, the command module forward heat shield is jettisoned at 24,000 feet, permitting mortar deployment of two reefed 16.5-foot diameter drogue parachutes for orienting and decelerating the spacecraft. After disreef and drogue release, three mortar deployed pilot chutes pull out the three main 83.3-foot diameter parachutes with two-stage reefing to provide gradual inflation in three steps. Two main parachutes out of three can provide a safe landing.

Reaction Control System (RCS) -- The SM RCS has four identical RCS "quads" mounted around the SM 90 degrees apart. Each quad has four 100 pound-thrust engines, two fuel and two oxidizer tanks and a helium pressurization sphere. Attitude control and small velocity maneuvers are made with the SM RCS.

The CM RCS consists of two independent six-engine subsystems of six 93 pound-thrust engines each used for spacecraft attitude control during entry. Propellants for both CM and SM RCS are monomethyl hydrazine fuel and nitrogen tetroxide oxidizer with helium pressurization. These propellants burn spontaneously when combined (without an igniter).

Electrical Power System (EPS) -- Provides electrical energy sources, power generation and control, power conversion, conditioning, and distribution to the spacecraft. The primary source of electrical power is the fuel cells mounted in the SM. The fuel cell also furnishes drinking water to the astronauts as a by-product.

Three silver-zinc oxide storage batteries supply power to the CM during entry and after landing, provide power for sequence controllers, and supplement the fuel cells during periods of peak power demand. A battery charger assures a full charge prior to entry.

Two other silver-zinc oxide batteries supply power for explosive devices for CM/SM separation, parachute deployment and separation, third-stage separation, launch escape tower separation, and other pyrotechnic uses.

Environmental Control System (ECS) -- Controls spacecraft atmosphere, pressure, and temperature and manages water. In addition to regulating cabin and suit gas pressure, temperature and humidity, the system removes carbon dioxide, odors and particles and ventilates the cabin after landing. It collects and stores fuel cell potable water for crew use, supplies water to the glycol evaporators for cooling, and dumps surplus water overboard through the waste H_2O dump nozzle. Proper operating temperature of electronics and electrical equipment is maintained by this system through the use of the cabin heat exchangers, the space radiators, and the glycol evaporators.

Recovery Aids -- Recovery aids include the uprighting system, swimmer interphone connections, sea dye marker, flashing beacon, VHF recovery beacon, and VHF transceiver. The uprighting system consists of three compressor-inflated bags to upright the spacecraft if it should land in the water apex down (stable II position).

Caution and Warning System -- Monitors spacecraft systems for out-of-tolerance conditions and alerts crew by visual and audible alarms.

Controls and Displays -- Provide status readouts and control functions of spacecraft systems in the command and service modules. All controls are designed to be operated by crewmen in pressurized suits. Displays are grouped by system and located according to the frequency of use and crew responsibility.

LUNAR MODULE STRUCTURES, WEIGHT

The lunar module is a two-stage vehicle designed for space operations near and on the Moon. The lunar module stands 22 feet 11 inches high and is 31 feet wide (diagonally across landing gear). The ascent and descent stages of the LM operate as a unit until staging, when the ascent stage functions as a single spacecraft for rendezvous and docking with the CM.

Ascent Stage

Three main sections make up the ascent stage: the crew compartment, midsection, and aft equipment bay. Only the crew compartment and midsection are pressurized (4.8 psig). The cabin volume is 235 cubic feet (6.7 cubic meters). The stage measures 12 feet 4 inches high by 14 feet 1 inch in diameter. The ascent stage has six substructural areas: crew compartment, midsection, aft equipment bay, thrust chamber assembly cluster supports, antenna supports, and thermal and micrometeoroid shield.

The cylindrical crew compartment is 92 inches (2.35 m) in diameter and 42 inches (1.07 m) deep. Two flight stations are equipped with control and display panels, armrests, body restraints, landing aids, two front windows, an overhead docking window, and an alignment optical telescope in the center between the two flight stations. The habitable volume is 160 cubic feet.

A tunnel ring atop the ascent stage meshes with the command module docking latch assemblies. During docking, the CM docking ring and latches are aligned by the LM drogue and the CSM probe.

The docking tunnel extends downward into the midsection 16 inches (40 cm). The tunnel is 32 inches (81 cm) in diameter and is used for crew transfer between the CSM and LM. The upper hatch on the inboard end of the docking tunnel opens inward and cannot be opened without equalizing pressure on both hatch surfaces.

A thermal and micrometeoroid shield of multiple layers of Mylar and a single thickness of thin aluminum skin encases the entire ascent stage structure.

DOCKING WINDOW

DOCKING DROGUE ASSEMBLY

VHF ANTENNA

DOCKING TARGET

EVA ANTENNA

S-BAND STEERABLE ANTENNA

RENDEZVOUS RADAR ANTENNA

S-BAND IN-FLIGHT ANTENNA (2)

AFT EQUIPMENT BAY

RCS THRUST CHAMBER ASSEMBLY CLUSTER (4)

WINDOWS (2)

TRACKING LIGHT

FORWARD HATCH

FORWARD

+Z

RCS PLUME DEFLECTORS

DOCKING LIGHT (4)

LANDING GEAR

RTG CASK

LANDING PAD

LADDER

EGRESS PLATFORM

DESCENT ENGINE SKIRT

LANDING RADAR ANTENNA

LUNAR SURFACE SENSING PROBE (3)

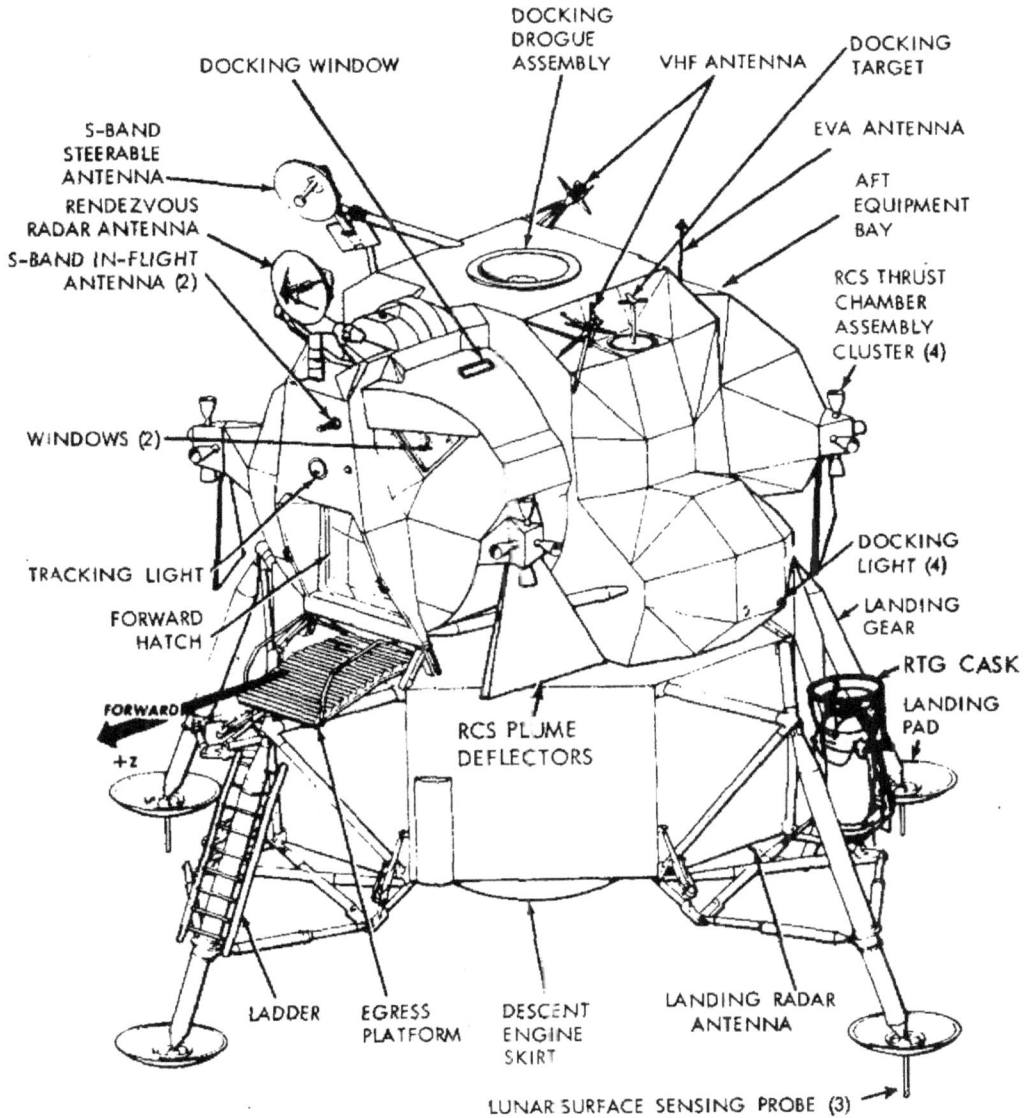

APOLLO LUNAR MODULE

-more-

Descent Stage

The descent stage center compartment houses the descent engine, and descent propellant tanks are housed in the four square bays around the engine. Quadrant II (Seq bay) contains ALSEP, and Radioisotope Thermoelectric Generator (RTG) externally. Quadrant IV contains the MESA. The descent stage measures 10 feet 7 inches high by 14 feet 1 inch in diameter and is encased in the Mylar and aluminum alloy thermal and micrometeoroid shield.

The LM egress platform, or "porch", is mounted on the forward outrigger just below the forward hatch. A ladder extends down the forward landing gear strut from the porch for crew lunar surface operations.

The landing gear struts are explosively extended and provide lunar surface landing impact attenuation. The main struts are filled with crushable aluminum honeycomb for absorbing compression loads. Footpads 37 inches (0.95 m) in diameter at the end of each landing gear provide vehicle support on the lunar surface.

Each pad (except forward pad) is fitted with a 68 inch long lunar surface sensing probe which signals the crew to shut down the descent engine upon contact with the lunar surface.

LM-7 flown on the Apollo 13 mission has a launch weight of 33,476 pounds. The weight breakdown is as follows:

Ascent stage, dry	4,668 lbs.	Includes water and oxygen; no crew
Descent stage, dry	4,650 lbs.	
RCS propellants (loaded)	590 lbs.	
DPS propellants (loaded)	18,339 lbs.	
APS propellants (loaded)	5,229 lbs.	
	33,476 lbs.	

Lunar Module Systems

Electrical Power System -- The LM DC electrical system consists of six silver zinc primary batteries -- four in the descent stage and two in the ascent stage. Twenty-eight-volt DC power is distributed to all LM systems. AC power (117v 400 Hz) is supplied by two inverters.

Environmental Control System -- Consists of the atmosphere
revitalization section, oxygen supply and cabin pressure control
section, water management, heat transport section, and outlets
for oxygen and water reservicing of the portable life support
system (PLSS).

Components of the atmosphere revitalization section are the
suit circuit assembly which cools and ventilates the pressure
garments, reduces carbon dioxide levels, removes odors, noxious
gases and excessive moisture; the cabin recirculation assembly
which ventilates and controls cabin atmosphere temperatures; and
the steam flex duct which vents to space steam from the suit
circuit water evaporator.

The oxygen supply and cabin pressure section supplies gaseous
oxygen to the atmosphere revitalization section for maintaining
suit and cabin pressure. The descent stage oxygen supply provides
descent flight phase and lunar stay oxygen needs, and the ascent
stage oxygen supply provides oxygen needs for the ascent and
rendezvous flight phase.

Water for drinking, cooling, fire fighting, food preparation,
and refilling the PLSS cooling water servicing tank is supplied by
the water management section. The water is contained in three
nitrogen-pressurized bladder-type tanks, one of 367-pound capacity
in the descent stage and two of 47.5-pound capacity in the ascent
stage.

The heat transport section has primary and secondary water-
glycol solution coolant loops. The primary coolant loop circulates
water-glycol for temperature control of cabin and suit circuit
oxygen and for thermal control of batteries and electronic components mounted on cold plates and rails. If the primary loop becomes
inoperative, the secondary loop circulates coolant through the
rails and cold plates only. Suit circuit cooling during secondary
coolant loop operation is provided by the suit loop water boiler.
Waste heat from both loops is vented overboard by water evaporation
or sublimators.

Communications System -- Two S-band transmitter-receivers,
two VHF transmitter-receivers, a signal processing assembly,
and associated spacecraft antenna make up the LM communications
system. The system transmits and receives voice and tracking and
ranging data, and transmits telemetry data on about 270 measurements and TV signals to the ground. Voice communications
between the LM and ground stations is by S-band, and between the
LM and CSM voice is on VHF.

Although no real-time commands can be sent to the LM, the digital uplink processes guidance officer commands, such as state vector updates, transmitted from Mission Control Center to the LM guidance computer.

The data storage electronics assembly (DSEA) is a four-channel voice recorder with timing signals, with a 10-hour recording capacity, which will be brought back into the CSM for return to Earth. DSEA recordings cannot be "dumped" to ground stations.

LM antennas are one 26-inch-diameter parabolic S-band steerable antenna, two S-band inflight antennas, two VHF inflight antennas, EVA antenna, and an erectable S-band antenna (optional) for lunar surface.

Guidance, Navigation, and Control System -- Comprised of six sections: primary guidance and navigation section (PGNS), abort guidance section (AGS), radar section, control electronics section (CES), and orbit rate display Earth and lunar (ORDEAL).

* The PGNS is an aided inertial guidance system updated by the alignment optical telescope, an inertial measurement unit, and the rendezvous and landing radars. The system provides inertial reference data for computations, produces inertial alignment reference by feeding optical sighting data into the LM guidance computer, displays position and velocity data, computes LM-CSM rendezvous data from radar inputs, controls attitude and thrust to maintain desired LM trajectory, and controls descent engine throttling and gimbaling.

The LM-7 primary guidance computer has the Luminary 1C Software program, which is an improved version over that in LM-6.

* The AGS is an independent backup system for the PGNS, having its own inertial sensors and computer.

* The radar section is made up of the rendezvous radar which provides CSM range and range rate, and line-of-sight angles for maneuver computation to the LM guidance computer; and the landing radar which provides altitude and velocity data to the LM guidance computer during lunar landing. The rendezvous radar has an operating range from 80 feet to 400 nautical miles. The ranging tone transfer assembly, utilizing VHF electronics, is a passive responder to the CSM VHF ranging device and is a backup to the rendezvous radar.

* The CES controls LM attitude and translation about all axes. It also controls by PGNS command the automatic operation of the ascent and descent engine and the reaction control thrusters. Manual attitude controller and thrust-translation controller commands are also handled by the CES.

*ORDEAL, displayed on the flight director attitude indicator, is the computed local vertical in the pitch axis during circular Earth or lunar orbits.

Reaction Control System -- The LM has four RCS engine clusters of four 100-pound (45.4 kg) thrust engines each, which use helium-pressurized hypergolic propellants. The oxidizer is nitrogen tetroxide, fuel is Aerozine 50 (50/50 blend of hydrazine and unsymmetrical dimethyl hydrazine). Interconnect valves permit the RCS system to draw from ascent engine propellant tanks.

The RCS provides small stabilizing impulses during ascent and descent burns, controls LM attitude during maneuvers, and produces thrust for separation, and for ascent/descent engine tank ullage. The system may be operated in either the pulse or steady-state modes.

Descent Propulsion System -- Maximum rated thrust of the descent engine is 9,870 pounds (4,380.9 kg) and is throttleable between 1,050 pounds (476.7 kg) and 6,300 pounds (2,860.2 kg). The engine can be gimbaled six degrees in any direction in response to attitude commands and to compensate for center of gravity offsets. Propellants are helium-pressurized Aerozine 50 and nitrogen tetroxide.

Ascent Propulsion System -- The 3,500-pound (1,589 kg) thrust ascent engine is not gimbaled and performs at full thrust. The engine remains dormant until after the ascent stage is separated from the descent stage. Propellants are the same as are burned by the RCS engines and the descent engine.

Caution and Warning, Controls and Displays -- These two systems have the same function aboard the lunar module as they do aboard the command module (See CSM systems section.)

Tracking and Docking Lights -- A flashing tracking light (once per second, 20 milliseconds duration) on the front face of the lunar module is an aid for contingency CSM-active rendezvous LM rescue. Visibility ranges from 400 nautical miles through the CSM sextant to 130 miles with the naked eye. Five docking lights analagous to aircraft running lights are mounted on the LM for CSM-active rendezvous: two forward yellow lights, aft white light, port red light and starboard green light. All docking lights have about a 1,000-foot visibility.

APOLLO 13 CREW AND CREW EQUIPMENT

Life Support Equipment - Space Suits

Apollo 13 crewmen will wear two versions of the Apollo space suit: an intravehicular pressure garment assembly worn by the command module pilot and the extravehicular pressure garment assembly worn by the commander and the lunar module pilot. Both versions are basically identical except that the extravehicular version has an integral thermal/meteoroid garment over the basic suit.

From the skin out, the basic pressure garment consists of a nomex comfort layer, a neoprene-coated nylon pressure bladder and a nylon restraint layer. The outer layers of the intravehicular suit are, from the inside out, nomex and two layers of Teflon-coated Beta cloth. The extravehicular integral thermal/meteoroid cover consists of a liner of two layers of neoprene-coated nylon, seven layers of Beta/Kapton spacer laminate, and an outer layer of Teflon-coated Beta fabric.

The extravehicular suit, together with a liquid cooling garment, portable life support system (PLSS), oxygen purge system, lunar extravehicular visor assembly and other components make up the extravehicular mobility unit (EMU). The EMU provides an extravehicular crewman with life support for a four-hour mission outside the lunar module without replenishing expendables. EMU total weight is 183 pounds. The intravehicular suit weighs 35.6 pounds.

Liquid cooling garment--A knitted nylon-spandex garment with a network of plastic tubing through which cooling water from the PLSS is circulated. It is worn next to the skin and replaces the constant wear-garment during EVA only.

Portable life support system--A backpack supplying oxygen at 3.9 psi and cooling water to the liquid cooling garment. Return oxygen is cleansed of solid and gas contaminants by a lithium hydroxide canister. The PLSS includes communications and telemetry equipment, displays and controls, and a main power supply. The PLSS is covered by a thermal insulation jacket. (Two stowed in LM).

Oxygen purge system--Mounted atop the PLSS, the oxygen purge system provides a contingency 45-minute supply of gaseous oxygen in two two-pound bottles pressurized to 5,880 psia. The system may also be worn separately on the front of the pressure garment assembly torso. It serves as a mount for the VHF antenna for the PLSS. (Two stowed in LM).

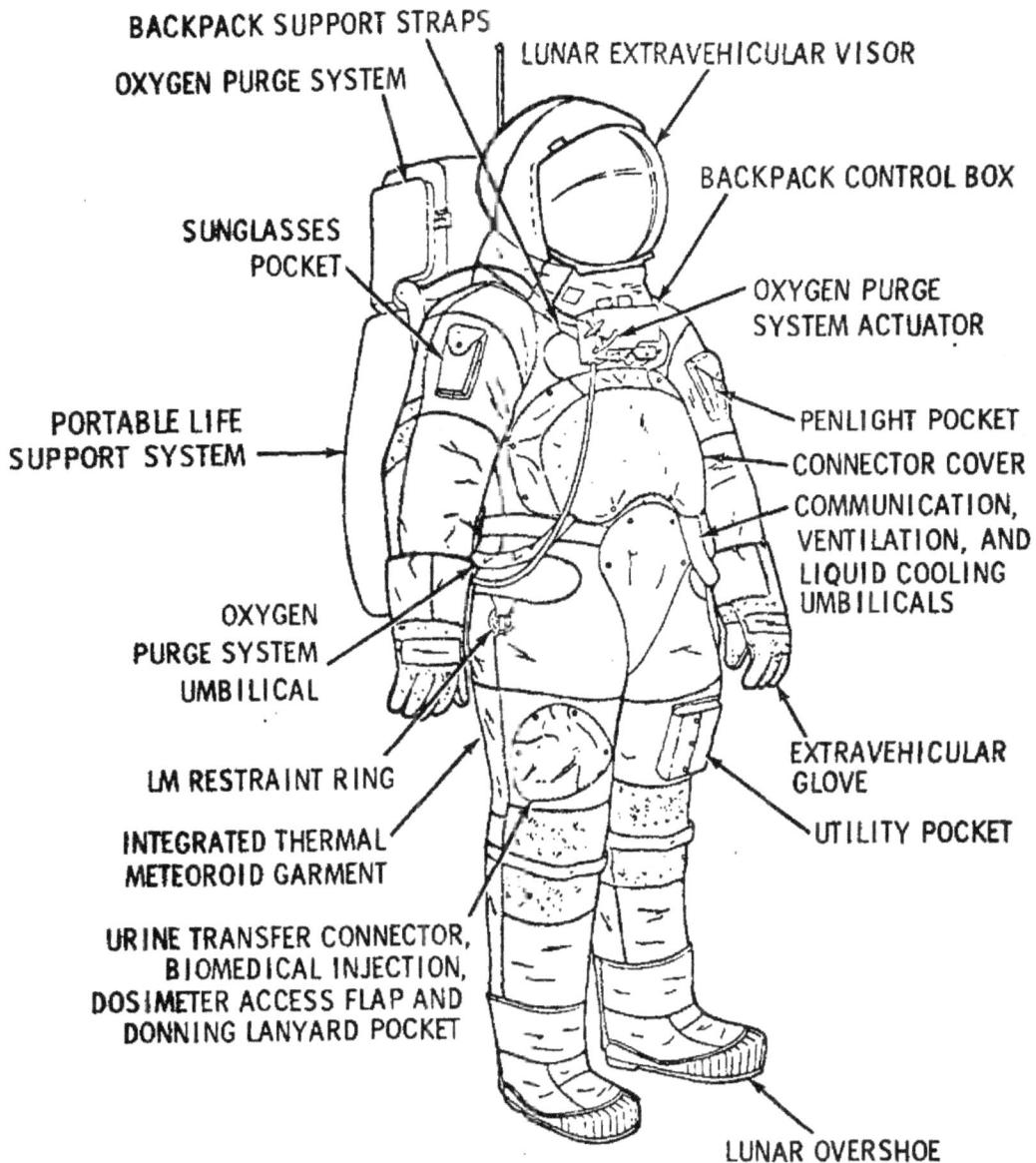

BACKPACK SUPPORT STRAPS

OXYGEN PURGE SYSTEM

LUNAR EXTRAVEHICULAR VISOR

BACKPACK CONTROL BOX

SUNGLASSES POCKET

OXYGEN PURGE SYSTEM ACTUATOR

PORTABLE LIFE SUPPORT SYSTEM

PENLIGHT POCKET

CONNECTOR COVER

COMMUNICATION, VENTILATION, AND LIQUID COOLING UMBILICALS

OXYGEN PURGE SYSTEM UMBILICAL

LM RESTRAINT RING

INTEGRATED THERMAL METEOROID GARMENT

EXTRAVEHICULAR GLOVE

UTILITY POCKET

URINE TRANSFER CONNECTOR, BIOMEDICAL INJECTION, DOSIMETER ACCESS FLAP AND DONNING LANYARD POCKET

LUNAR OVERSHOE

-more-

EXTRAVEHICULAR MOBILITY UNIT

Lunar extravehicular visor assembly--A polycarbonate
shell and two visors with thermal control and optical
coatings on them. The EVA visor is attached over the
pressure helmet to provide impact, micrometeoroid, thermal
and ultraviolet-infrared light protection to the EVA crewmen. Since Apollo 12, a sunshade has been added to the outer
portion of the LEVVA in the middle portion of the helmet rim.

Extravehicluar gloves--Built of an outer shell of
Chromel-R fabric and thermal insulation to provide protection
when handling extremely hot and cold objects. The finger
tips are made of silicone rubber to provide more sensitivity.

A one-piece constant-wear garment, similar to "long
johns," is worn as an undergarment for the space suit in intravehicular operations and for the inflight coveralls. The
garment is porous-knit cotton with a waist-to-neck zipper for
donning. Biomedical harness attach points are provided.

During periods out of the space suits, crewmen wear two-
piece Teflon fabric inflight coveralls for warmth and for
pocket stowage of personal items.

Communications carriers ("Snoopy Hats") with redundant
microphones and earphones are worn with the pressure helmet;
a lightweight headset is worn with the inflight coveralls.

Another modification since Apollo 12 has been the addition
of eight-ounce drinking water bags ("Gunga Dins") attached to
the inside neck rings of the EVA suits. The crewmen can take
a sip of water from the 6 X 8 inch bag through a 1/8-inch-
diameter tube within reach of his mouth. The bags are filled
from the lunar module potable water dispenser.

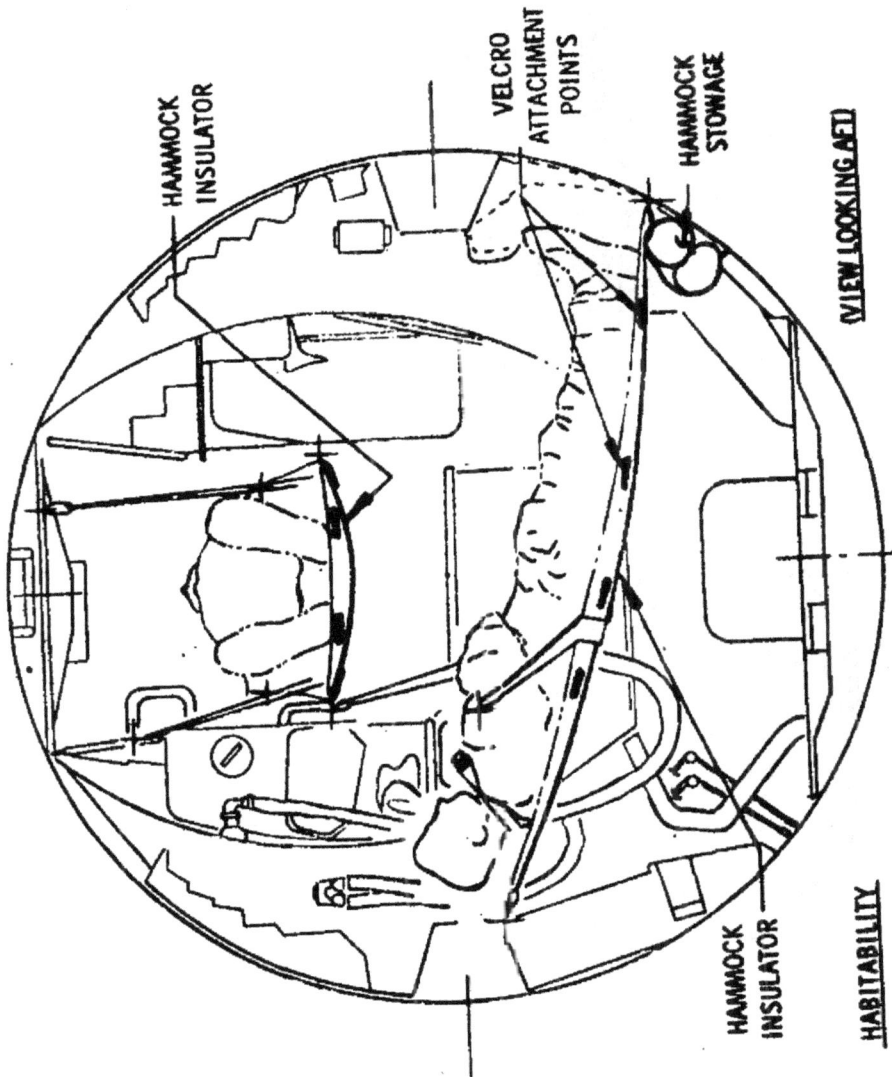

HAMMOCK
INSULATOR

VELCRO
ATTACHMENT
POINTS

HAMMOCK
STOWAGE

(VIEW LOOKING AFT)

HAMMOCK
INSULATOR

HABITABILITY

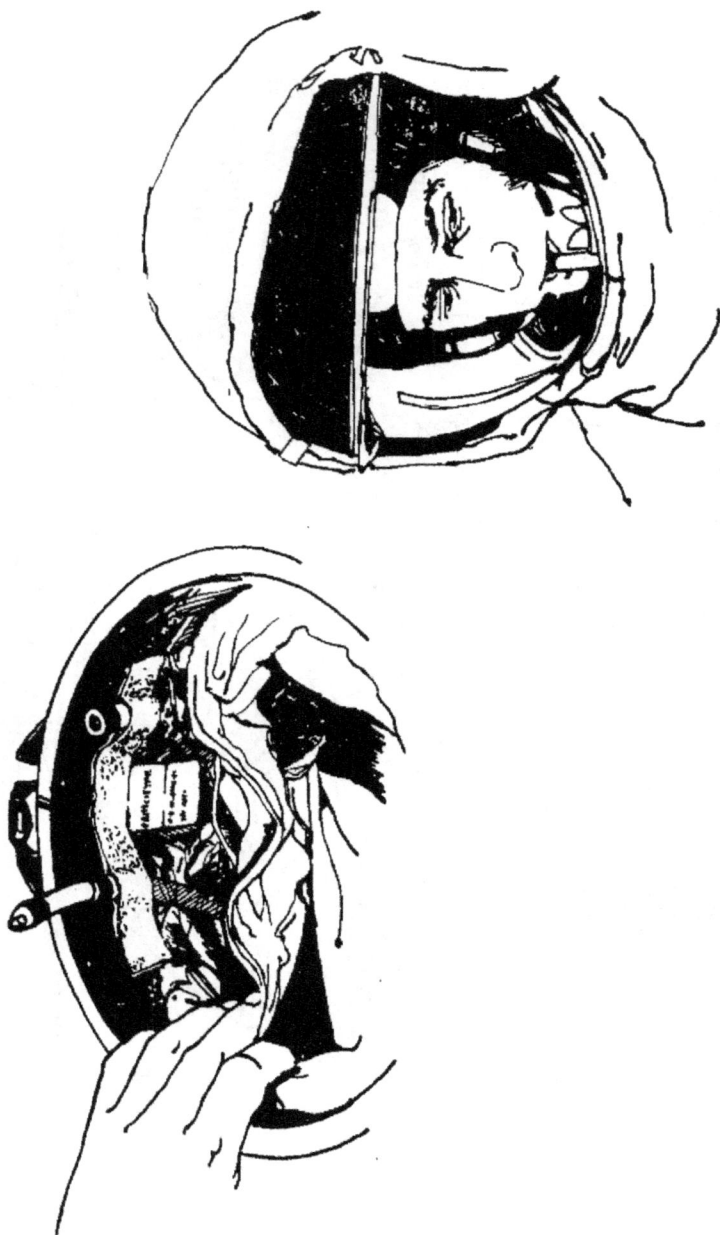

WATER BAG

APOLLO 13 SPACESUIT WATER BAG FOR USE DURING EVA.

Apollo Lunar Hand Tools

Special Environmental Container - The special environmental sample is collected in a carefully selected area and sealed in a special container which will retain a high vacuum. The container is opened in the Lunar Receiving Laboratory where it will provide scientists the opportunity to study lunar material in its original environment.

Extension handle - This tool is of aluminum alloy tubing with a malleable stainless steel cap designed to be used as an anvil surface. The handle is designed to be used as an extension for several other tools and to permit their use without requiring the astronaut to kneel or bend down. The handle is approximately 24 inches long and 1 inch in diameter. The handle contains the female half of a quick disconnect fitting designed to resist compression, tension, torsion, or a combination of these loads.

Three core tubes - These tubes are designed to be driven or augered into loose gravel, sandy material, or into soft rock such as feather rock or pumice. They are about 15 inches in length and one inch in diameter and are made of aluminum tubing. Each tube is supplied with a removeable non-serrated cutting edge and a screw-on cap incorporating a metal-to-metal crush seal which replaces the cutting edge. The upper end of each tube is sealed and designed to be used with the extension handle or as an anvil. Incorporated into each tube is a spring device to retain loose materials in the tube.

Scoops (large and small) - These tools are designed for use as a trowel and as a chisel. The scoop is fabricated primarily of aluminum with a hardened-steel cutting edge riveted on and a nine-inch handle. A malleable stainless steel anvil is on the end of the handle. The angle between the scoop pan and the handle allows a compromise for the dual use. The scoop is used either by itself or with the extension handle. The large scoop has a seive which permits particles smaller than 1/2 cm to pass through.

Sampling hammer - This tool serves three functions, as a sampling hammer, as a pick or mattock, and as a hammer to drive the core tubes or scoop. The head has a small hammer face on one end, a broad horizontal blade on the other, and large hammering flats on the sides. The handle is 14 inches long and is made of formed tubular aluminum. The hammer has on its lower end a quick-disconnect to allow attachment to the extension handle for use as a hoe. The head weight has been increased to provide more impact force.

BRUSH / SCRIBER / HAND LENS

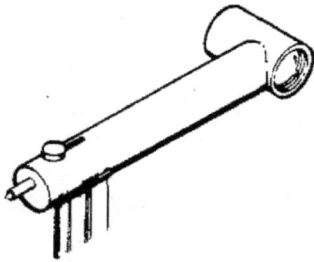

SPRING SCALE

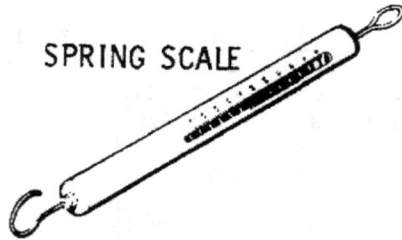

SCOOP

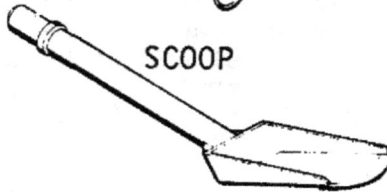

HAMMER

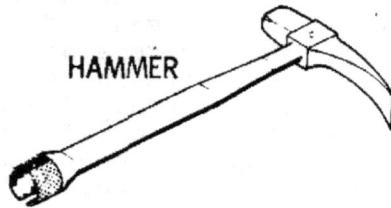

CORE TUBE
AND CAP

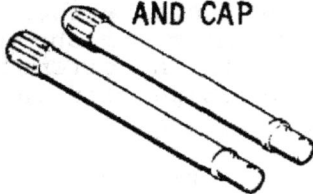

TONGS

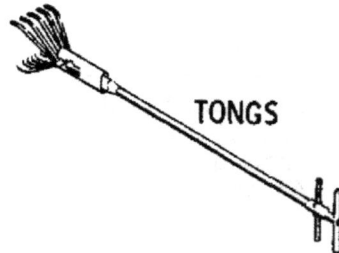

GEOLOGIC SAMPLING TOOLS

-more-

Tongs - The tongs are designed to allow the astronaut to retrieve small samples from the lunar surface while in a standing position. The tines are of such angles, length, and number to allow samples of from 3/8 up to 2-1/2-inch diameter to be picked up. This tool is 24 inches in overall length.

Brush/Scriber/Hand Lens - A composite tool

(1) Brush - To clean samples prior to selection
(2) Scriber - To scratch samples for selection and to mark for identification
(3) Hand lens - Magnifying glass to facilitate sample selection

Spring Scale - To weigh two rock boxes and other bags containing lunar material samples, to maintain weight budget for return to Earth.

Instrument staff - The staff hold the Hasselblad camera. The staff breaks down into sections. The upper section telescopes to allow generation of a vertical stereoscopic base of one foot for photography. Positive stops are provided at the extreme of travel. A shaped hand grip aids in aiming and carrying. The bottom section is available in several lengths to suit the staff to astronauts of varying sizes. The device is fabricated from tubular aluminum.

Gnomon - This tool consists of a weighted staff suspended on a two-ring gimbal and supported by a tripod. The staff extends 12 inches above the gimbal and is painted with a gray scale. The gnomon is used as a photographic reference to indicate local vertical, sun angle, and scale. The gnomon has a required accuracy of vertical indication of 20 minutes of arc. Magnetic damping is incorporated to reduce oscillations.

Color Chart - The color chart is painted with three primary colors and a gray scale. It is used as a calibration for lunar photography. The scale is mounted on the tool carrier but may easily be removed and returned to Earth for reference. The color chart is 6 inches in size.

Tool Carrier - The carrier is the stowage container for
the tools during the lunar flight. After the landing the
carrier serves as support for the astronaut when he kneels
down, as a support for the sample bags and samples, and as
a tripod base for the instrument staff. The carrier folds
flat for stowage. For field use it opens into a triangular
configuration. The carrier is constructed of formed sheet
metal and approximates a truss structure. Six-inch legs
extend from the carrier to elevate the carrying handle suffi-
ciently to be easily grasped by the astronaut.

Field Sample Bags - Approximately 80 bags four inches by
five inches are included in the Apollo lunar hand tools for
the packaging of samples. These bags are fabricated from
Teflon FEP.

Collection Bag - This is a large bag (4 X 8 inches)
attached to the astronaut's side of the tool carrier. Field
sample bags are stowed in this bag after they have been filled.
It can also be used for general storage or to hold items
temporarily. (Two in each SRC).

Trenching Tool - A trenching tool with a pivoting scoop
has been provided for digging the two-foot deep soil mechanics
investigation trench. The two-piece handle is five feet long.
The scoop is eight inches long and five inches wide and pivots
from in-line with the handle to 90°--similar to the trenching
tool carried on infantry backpacks. The trenching tool is
stowed in the MESA rather than in the tool carrier.

Lunar Surface Drill - The 29.4-pound Apollo Lunar Surface
Drill (ALSD) is stowed in the ALSEP subpackage No. 2 and will
be used for boring two ten-foot deep 1.25-inch diameter holes
for ALSEP heat flow experiment probes, and one approximately
eight-foot-deep, one-inch-diameter core sample. The silver-
zinc battery-powered rotary percussive drill has a clutch to
limit torque to 20 foot-pounds. A treadle assembly serves as
a drilling platform and as a core stem lock during the drill
string decoupling operation as the string is withdrawn from
the lunar soil. Bore stems for the heat flow experiment holes
are of boron/fiberglas, and the core sample core stems are
titanium. Cutting bits are tungsten carbide.

Apollo 13 Crew Menu

More than 70 items comprise the food selection list of freeze-dried rehydratable, wet-pack and spoon-bowl foods. Balanced meals for five days have been packed in man/day overwraps. Items similar to those in the daily menus have been packed in a snack pantry. The snack pantry permits the crew to locate easily a food item in a smorgasbord mode without having to "rob" a regular meal somewhere down deep in a storage box.

Water for drinking and rehydrating food is obtained from two sources in the command module--a dispenser for drinking water and a water spigot at the food preparation station supplying water at about 155 or 55° F. The potable water dispenser squirts water continuously as long as the trigger is held down, and the food preparation spigot dispenses water in one-ounce increments.

A continuous-feed hand water dispenser similar to the one in the command module is used aboard the lunar module for cold-water rehydration of food packets stowed aboard the LM.

After water has been injected into a food bag, it is kneaded for about three minutes. The bag neck is then cut off and the food squeezed into the crewman's mouth. After a meal, germicide pills attached to the outside of the food bags are placed in the bags to prevent fermentation and gas formation. The bags are then rolled and stowed in waste disposal compartments.

The day-by-day, meal-by-meal Apollo 13 Menu for Commander Lovell is on the following page as a typical five-day menu for each crewman.

TYPICAL CREW MENU IS THAT OF APOLLO 13 COMMANDER LOVELL:

MEAL	Day 1*, 5**, 9	Day 2, 6, 10	Day 3, 7, 11	Day 4, 8
A.	Peaches RSB Canadian Bacon & Applesauce RSB Bacon Squares (8) IMB Cocoa R Orange Drink R	Pears IMB Bacon Squares (8) IMB Scrambled Eggs RSB Grapefruit Drink R Coffee (b) R	Peaches IMB Canadian Bacon & Applesauce RSB Sugar Coated Corn Flakes RSB Cocoa R Grape Drink R	Apricots IMB Bacon Squares (8) IMB Scrambled Eggs RSB Orange-G.F. Drink R Coffee(B) R
B.	Salmon Salad RSB Beef & Gravy WP Jellied Candy IMB Grape Drink R	Frankfurters WP Cranberry-Orange RSB Chocolate Pudding RSB Orange-G.F. Drink R	Cream of Chicken Soup RSB Bread Slice *** Sandwich Spread WP Chocolate Bar IMB P.A.-G.F. Drink R	Chicken & Rice Soup RSB Meatballs with Sauce WP Caramel Candy IMB Orange Drink R
C.	Pea Soup RSB Chicken & Rice RSB Date Fruitcake (4) DB P.A.-G.F. Drink R	Shrimp Cocktail RSB Pork & Scalloped Potatoes RSB Apricots IMB Orange Drink R	Chicken Stew RSB Turkey & Gravy WP Butterscotch Pudding RSB Grapefruit Drink R	Tuna Salad RSB Beef Stew RSB Banana Pudding RSB Grape Punch R
TOTAL CALORIES	2106	2073	2183	2043

DB = Dry Bite
IMB = Intermediate Moisture Bite
R = Rehydratable
RSB = Rehydratable Spoon Bowl
WP = Wet Pack

* Day 1 consists of Meal B and C only; extra meal consists of: Ham & Cheese Sandwich (frozen), Caramel Candy, Orange-G.F. Drink
** Day 5 consists of Meal A only
*** Bread: Cheese, Rye, White
   Sandwich Spreads: Chicken, Ham, Tuna Salad, Cheddar Cheese Spread, Peanut Butter, Jelly.

Personal Hygiene

Crew personal hygiene equipment aboard Apollo 13
includes body cleanliness items, the waste management system
and one medical kit.

Packaged with the food are a toothbrush and a two-ounce
tube of toothpaste for each crewman. Each man-meal package
contains a 3.5-by-4-inch wet-wipe cleansing towel. Additionally, three packages of 12-by-12-inch dry towels are stowed
beneath the command module pilot's couch. Each package contains seven towels. Also stowed under the command module
pilot's couch are seven tissue dispensers containing 53 three-
ply tissues each.

Solid body wastes are collected in plastic defecation
bags which contain a germicide to prevent bacteria and gas
formation. The bags are sealed after use and stowed in empty
food containers for post-flight analysis.

Urine collection devices are provided for use while
wearing either the pressure suit or the inflight coveralls.
The urine is dumped overboard through the spacecraft urine dump
valve in the CM and stored in the LM.

Medical Kit

The 5X5X8-inch medical accessory kit is stowed in a compartment on the spacecraft right side wall beside the lunar
module pilot couch. The medical kit contains three motion
sickness injectors, three pain suppression injectors, one two-
ounce bottle first aid ointment, two one-ounce bottles eye
drops, three nasal sprays, two compress bandages, 12 adhesive
bandages, one oral thermometer, and four spare crew biomedical
harnesses. Pills in the medical kit are 60 antibiotic, 12
nausea, 12 stimulant, 18 pain killer, 60 decongestant, 24
diarrhea, 72 aspirin and 21 sleeping. Additionally, a small
medical kit containing four stimulant, eight diarrhea, two
sleeping and four pain killer pills, 12 aspirin, one bottle
eye drops, two compress bandages, 8 decongestant pills, one
pain injector, one bottle nasal spray is stowed in the lunar
module flight data file compartment.

-more-

Survival Gear

The survival kit is stowed in two rucksacks in the right-hand forward equipment bay above the lunar module pilot.

Contents of rucksack No. 1 are: two combination survival lights, one desalter kit, three pair sunglasses, one radio beacon, one spare radio beacon battery and spacecraft connector cable, one knife in sheath, three water containers, and two containers of Sun lotion, two utility knives, three survival blankets and one utility netting.

Rucksack No. 2: one three-man life raft with CO_2 inflater, one sea anchor, two sea dye markers, three sunbonnets, one mooring lanyard, three manlines and two attach brackets.

The survival kit is designed to provide a 48-hour post-landing (water or land) survival capability for three crewmen between 40° North and South latitudes.

Biomedical Inflight Monitoring

The Apollo 13 crew biomedical telemetry data received by the Manned Space Flight Network will be relayed for instantaneous display at Mission Control Center where heart rate and breathing rate data will be displayed on the flight surgeon's console. Heart rate and respiration rate average, range and deviation are computed and displayed on digital TV screens.

In addition, the instantaneous heart rate, real-time and delayed EKG and respiration are recorded on strip charts for each man.

Biomedical telemetry will be simultaneous from all crewmen while in the CSM, but selectable by a manual onboard switch in the LM.

Biomedical data observed by the flight surgeon and his team in the Life Support Systems Staff Support Room will be correlated with spacecraft and space suit environmental data displays.

Blood pressures are no longer telemetered as they were in the Mercury and Gemini programs. Oral temperatures, however, can be measured onboard for diagnostic purposes and voiced down by the crew in case of inflight illness.

Energy expended by the crewmen during EVA will be
determined indirectly using a metabolic computation program based on three separate measurements:

1) Heart rate portion -- Heart rate will be determined
   from telemetered EKG and converted to oxygen con-
   sumption (litre/min) and heat production (BTU/hour)
   based on pre-flight calibration curves. These curves
   are determined from exercise response tests utilizing
   a bicycle ergometer.

2) Oxygen usage portion -- Oxygen usage will be determined
   from the telemetered measurement of PLSS oxygen supply
   pressure. Suit leak determined pre-flight is taken
   into account. Heat production will be calculated from
   oxygen usage.

3) Liquid cooled garment temperature portion -- The
   amount of heat taken up by the liquid cooled garment
   will be determined from telemetered measurements of
   the LCG water temperature inlet and change in/out.
   This measurement (the amount of heat taken up by the
   water) plus an allowance made for sensible and latent
   heat loss, radiant heat load, and possible heat storage
   will provide an indication of heat production by the
   crewman.

Training

The crewmen of Apollo 13 have spent more than five hours of formal crew training for each hour of the lunar-launching mission's ten-day duration. More than 1,000 hours of training were in Apollo 13 crew training syllabus over and above the normal preparations for the mission--technical briefings and reviews, pilot meetings and study.

The Apollo 13 crewmen also took part in prelaunch testing at Kennedy Space Center, such as altitude chamber tests and the countdown demonstration tests (CDDT) which provided the crew with thorough operational knowledge of the complex vehicle.

Highlights of specialized Apollo 13 crew training topics are:

* Detailed series of briefings on spacecraft systems, operation and modifications.

* Saturn launch vehicle briefings on countdown, range safety, flight dynamics, failure modes and abort conditions. The launch vehicle briefings were updated periodically.

* Apollo Guidance and Navigation system briefings at the Massachusetts Institute of Technology Instrumentation Laboratory.

* Briefings and continuous training on mission photographic objectives and use of camera equipment.

* Extensive pilot participation in reviews of all flight procedures for normal as well as emergency situations.

* Stowage reviews and practice in training sessions in the spacecraft, mockups and command module simulators allowed the crewmen to evaluate spacecraft stowage of crew-associated equipment.

* More than 400 hours of training per man in command module and lunar module simulators at MSC and KSC, including closed-loop simulations with flight controllers in the Mission Control Center. Other Apollo simulators at various locations were used extensively for specialized crew training.

* Lunar surface briefings and some 20 suited 1-g walk-
throughs of lunar surface EVA operations covering lunar
geology and microbiology and deployment of experiments in
the Apollo Lunar Surface Experiment Package (ALSEP). Training in lunar surface EVA included practice sessions with lunar
surface sample gathering tools and return containers, cameras,
the erectable S-band antenna and the modular equipment stowage
assembly (MESA) housed in the LM descent stage.

* Proficiency flights in the lunar landing training
vehicle (LLTV) for the commander.

* Zero-g and one-sixth g aircraft flights using command
module and lunar module mockups for EVA and pressure suit
doffing/donning practice and training.

* Underwater zero-g training in the MSC Water Immersion
Facility using spacecraft mockups to further familiarize
crew with all aspects of CSM-LM docking tunnel intravehicular
transfer and EVA in pressurized suits.

* Water egress training conducted in indoor tanks as
well as in the Gulf of Mexico, included uprighting from the
Stable II position (apex down) to the Stable I position (apex
up), egress onto rafts donning Biological Isolation Garments
(BIGs), decontamination procedures and helicopter pickup.

* Launch pad egress training from mockups and from the
actual spacecraft on the launch pad for possible emergencies
such as fire, contaminants and power failures.

* The training covered use of Apollo spacecraft fire
suppression equipment in the cockpit.

* Planetarium reviews at Morehead Planetarium, Chapel
Hill, N.C., and at Griffith Planetarium, Los Angeles, Calif.,
of the celestial sphere with special emphasis on the 37
navigational stars used by the Apollo guidance computer.

NATIONAL AERONAUTICS AND SPACE ADMINISTRATION

WASHINGTON, D. C. 20544

BIOGRAPHICAL DATA

NAME: James Arthur Lovell, Jr. (Captain, USN)
NASA Astronaut

BIRTHPLACE AND DATE: Born March 25, 1928, in Cleveland, Ohio.
His mother, Mrs. Blanche Lovell, resides at Edgewater
Beach, Florida.

PHYSICAL DESCRIPTION: Blond hair; blue eyes; height: 5 feet
11 inches; weight: 170 pounds.

EDUCATION: Graduated from Juneau High School, Milwaukee,
Wisconsin; attended the University of Wisconsin for
2 years, then received a Bachelor of Science degree
from the United States Naval Academy in 1952; presented
an Honorary Doctorate from Illinois Wesleyan University
in 1969.

MARITAL STATUS: Married to the former Marilyn Gerlach of
Milwaukee, Wisconsin. Her parents, Mr. and Mrs. Carl
Gerlach, are residents of Milwaukee.

CHILDREN: Barbara L., October 13, 1953; James A., February
15, 1955; Susan K., July 14, 1958; Jeffrey C.,
January 14, 1966.

RECREATIONAL INTERESTS: His hobbies are golf, swimming,
handball, and tennis.

ORGANIZATIONS: Member of the Society of Experimental Test
Pilots and the Explorers Club.

SPECIAL HONORS: Awarded the NASA Distinguished Service Medal,
two NASA Exceptional Service Medals, the Navy Astronaut
Wings, the Navy Distinguished Service Medal, and two
Navy Distinguished Flying Crosses; recipient of the
1967 FAI De Laval and Gold Space Medals (Athens, Greece),
the American Academy of Achievement Golden Plate Award,
the City of New York Gold Medal in 1969, the City of
Houston Medal for Valor in 1969, the National Geographic
Society's Hubbard Medal in 1969, the National Academy
of Television Arts and Sciences Special Trustees Award
in 1969, and the Institute of Navigation Award in 1969.

Co-recipient of the American Astronautical Society Flight
Achievement Awards in 1966 and 1968, the Harmon International Trophy in 1966 and 1967, the Robert H. Goddard
Memorial Trophy in 1969, the H. H. Arnold Trophy for
1969, the General Thomas D. White USAF Space Trophy for
1968, the Robert J. Collier Trophy for 1968, and the
1969 Henry G. Bennett Distinguished Service Award.

EXPERIENCE: Lovell, a Navy Captain, received flight training
following graduation from Annapolis in 1952.

He has had numerous naval aviator assignments including
a 4-year tour as a test pilot at the Naval Air Test
Center, Patuxent River, Maryland. While there he served
as program manager for the F4H weapon system evaluation.
A graduate of the Aviation Safety School of the University of Southern California, he also served as a flight
instructor and safety engineer with Fighter Squadron
101 at the Naval Air Station, Oceana, Virginia.

He has logged more than 4,407 hours flying time--more than
3,000 hours in jet aircraft.

CURRENT ASSIGNMENT: Captain Lovell was selected as an astronaut
by NASA in September 1962. He has since served as backup pilot for the Gemini 4 flight and backup command pilot
for the Gemini 9 flight.

On December 4, 1965, he and Command pilot Frank Borman were
launched into space on the history-making Gemini 7 mission.
The flight lasted 330 hours and 35 minutes, during which
the following space firsts were accomplished: longest
manned space flight; first rendezvous of two manned
maneuverable spacecraft, as Gemini 7 was joined in orbit
by Gemini 6; and longest multi-manned space flight. It
was also on this flight that numerous technical and
medical experiments were completed successfully.

The Gemini 12 mission, with Lovell and pilot Edwin Aldrin,
began on November 11, 1966. This 4-day, 59-revolution
flight brought the Gemini Program to a successful close.
Major accomplishements of the 94-hour 35-minute flight
included a third-revolution rendezvous with the previously
launched Agena (using for the first time backup onboard
computations due to radar failure); a tethered station-
keeping exercise,; retrieval of a micrometeorite experiment package from the spacecraft exterior; an evaluation
of the use of body restraints specially designed for
completing work tasks outside of the spacecraft; and
completion of numerous photographic experiments, highlights of which are the first pictures taken from space
of an eclipse of the sun.

-more-

Gemini 12 ended with retrofire at the beginning of the 60th revolution, followed by the second consecutive fully automatic controlled reentry of a spacecraft, and a landing in the Atlantic within 2 1/2 miles of the USS WASP.

As a result of his participation in the Gemini 7 and 12 flights, Lovell logged 425 hours and 10 minutes in space. Aldrin established a new EVA record by completing 5 1/2 hours outside the spacecraft during two standup EVAs and one umbilical EVA.

Lovell served as command module pilot for the epic six-day journey of Apollo 8--man's maiden voyage to the moon --December 21-27, 1968. Apollo 8 was the first manned spacecraft to be lifted into near-earth orbit by a 7 1/2-million pound thrust Saturn V launch vehicle, and all events in the flight plan occurred as scheduled with unbelievable accuracy.

A "go" for the translunar injection burn was given midway through the second near-earth orbit, and the restart of the S-IVB third stage to effect this maneuver increased the spacecraft's velocity to place it on an intercept course with the moon. Lovell and fellow crew members, Frank Borman (spacecraft commander) and William A. Anders (lunar module pilot), piloted their spacecraft some 223,000 miles to become the first humans to leave the earth's influence; and upon reaching the moon on December 24, they performed the first critical maneuver to place Apollo 8 into a 60 by 168 nautical miles lunar orbit.

Two revolutions later, the crew executed a second maneuver using the spacecraft's 20,500-pound thrust service module propulsion system to achieve a circular lunar orbit of 60 nautical miles. During their ten revolutions of the moon, the crew conducted live television transmissions of the lunar surface and performed such tasks as landmark and Apollo landing site tracking, vertical stereo photography and stereo navigation photography, and sextant navigation using lunar landmarks and stars. At the end of the tenth lunar orbit, they executed a transearth injection burn which placed Apollo 8 on a proper trajectory for the return to earth.

-more-

The final leg of the trip required only 58 hours, as compared to the 69 hours used to travel to the moon, and Apollo 8 came to a successful conclusion on December 27, 1968. Splashdown occurred at an estimated 5,000 yeards from the USS YORKTOWN, following the successful negotiation of a critical 28-mile high reentry corridor at speeds close to 25,000 miles per hour.

Captain Lovell has since served as the backup spacecraft commander for the Apollo 11 lunar landing mission. He has completed three space flights and holds the U.S. Astronaut record for time in space with a total of 572 hours and 10 minutes.

SPECIAL ASSIGNMENT: In addition to his regular duties as an astronaut, Captain Lovell continues to serve as Special Consultant to the President's Council on Physical Fitness and Sports--an assignment he has held since June 1967.

CURRENT SALARY: $1,717.28 per month.

-more-

NATIONAL AERONAUTICS AND SPACE ADMINISTRATION

WASHINGTON, D. C. 20546

BIOGRAPHICAL DATA

NAME: Thomas Kenneth Mattingly II (Lieutenant Commander, USN)
NASA Astronaut

BIRTHPLACE AND DATE: Born in Chicago, Ill., March 17, 1936.
His parents, Mr. and Mrs. Thomas K. Mattingly, now
reside in Hialeah, Fla.

PHYSICAL DESCRIPTION: Brown hair; blue eyes; height: 5 feet
10 inches; weight: 140 pounds.

EDUCATION: Attended Florida elementary and secondary
schools and is a graduate of Miami Edison High School,
Miami, Fla.; received a Bachelor of Science degree in
Aeronautical Engineering from Auburn University in
1958.

MARITAL STATUS: Single

RECREATIONAL INTERESTS: Enjoys water skiing and playing
handball and tennis.

ORGANIZATIONS: Member of the American Institute of Aeronautics and Astronautics and the U.S. Naval Institute.

EXPERIENCE: Prior to reporting for duty at the Manned
Spacecraft Center, he was a student at the Air Force
Aerospace Research Pilot School.

He began his Naval career as an Ensign in 1958 and
received his wings in 1960. He was then assigned to
VA-35 and flew A1H aircraft aboard the USS SARATOGA
from 1960 to 1963. In July 1963, he served in VAH-11
deployed aboard the USS FRANKLIN D. ROOSEVELT where
he flew the A3B aircraft for two years.

He has logged 3,700 hours of flight time--1,946 hours
in jet aircraft.

CURRENT ASSIGNMENT: Lt Commander Mattingly is one of the 19
astronauts selected by NASA in April 1966. He served
as a member of the astronaut support crews for the
Apollo 8 and 11 missions.

CURRENT SALARY: $1,293.33 per month.

more

NATIONAL AERONAUTICS AND SPACE ADMINISTRATION

WASHINGTON, D. C. 20546

BIOGRAPHICAL DATA

NAME: Fred Wallace Haise, Jr. (Mr.)
NASA Astronaut

BIRTHPLACE AND DATE: Born in Biloxi, Miss., on Nov. 14, 1933; his mother, Mrs. Fred W. Haise, Sr., resides in Biloxi.

PHYSICAL DESCRIPTION: Brown hair; brown eyes; height: 5 feet 9 1/2 inches; weight: 150 pounds.

EDUCATION: Graduated from Biloxi High School, Biloxi, Miss.; attended Perkinston Junior College (Association of Arts); received a Bachelor of Science degree with honors in Aeronautical Engineering from the University of Oklahoma in 1959.

MARITAL STATUS: Married to the former Mary Griffin Grant of Biloxi, Miss. Her parents, Mr. and Mrs. William J. Grant, Jr., reside in Biloxi.

CHILDREN: Mary M., January 25, 1956; Frederick T., May 13, 1958; Stephen W., June 30, 1961.

ORGANIZATIONS: Member of the Society of Experimental Test Pilots, Tau Beta Pi, Sigma Gamma Tau, and Phi Theta Kappa.

SPECIAL HONORS: Recipient of the A. B. Honts Trophy as the outstanding graduate of class 64A from the Aerospace Research Pilot School in 1964; awarded the American Defense Ribbon and the Society of Experimental Test Pilots Ray E. Tenhoff Award for 1966.

EXPERIENCE: Haise was a research pilot at the NASA Flight Research Center at Edwards, Calif., before coming to Houston and the Manned Spacecraft Center; and from September 1959 to March 1963, he was a research pilot at the NASA Lewis Research Center in Cleveland, Ohio. During this time, he authored the following papers which have been published: a NASA TND, entitled "An Evaluation of the Flying Qualities of Seven General-Aviation Aircraft;" NASA TND 3380, "Use of Aircraft for Zero Gravity Environment, May 1966;" SAE Business Aircraft Conference Paper, entitled "An Evaluation of General-Aviation Aircraft Flying Qualities," March 30-April 1, 1966; and a paper delivered at the tenth symposium of the Society of

Experimental Test Pilots, entitled "A Quantitative/
Qualitative Handling Qualities Evaluation of Seven
General-Aviation Aircraft," 1966.

He was the Aerospace Research Pilots School's outstanding graduate of Class 64A and served with the
U.S. Air Force from October 1961 to August 1962 as a
tactical fighter pilot and as Chief of the 164th
Standardization-Evaluation Flight of the 164th Tactical
Fighter Squadron at Mansfield, Ohio. From March 1957
to September 1959, he was a fighter-interceptor pilot
with the 185th Fighter Interceptor Squadron in the Oklahoma Air National Guard.

He also served as a tactics and all weather flight
instructor in the U.S. Navy Advanced Training Command
at NAAS Kingsville, Texas, and was assigned as a U.S.
Marine Corps fighter pilot to VMF-533 and 114 at MCAS
Cherry Point, N.C., from March 1954 to September 1956.

His military career began in October 1952 as a Naval
Aviation Cadet at the Naval Air Station in Pensacola,
Fla.

He has accumulated 5,800 hours flying time, including
3,000 hours in jets.

CURRENT ASSIGNMENT: Mr. Haise is one of the 19 astronauts
selected by NASA in April 1966. He served as backup
lunar module pilot for the Apollo 8 and 11 missions.

CURRENT SALARY: $1,698.00 per month.

LAUNCH COMPLEX 39

Launch Complex 39 facilities at the Kennedy Space Center were planned and built specifically for the Apollo Saturn V, the space vehicle being used in the United States manned lunar exploration program.

Complex 39 introduced the mobile concept of launch operations in which the space vehicle is thoroughly checked out in an enclosed building before it is moved to the launch pad for final preparations. This affords greater protection from the elements and permits a high launch rate since pad time is minimal.

Saturn V stages are shipped to the Kennedy Space Center by ocean-going vessels and specially designed aircraft. Apollo spacecraft modules are transported by air and first taken to the Manned Spacecraft Operations Building in the Industrial Area south of Complex 39 for preliminary checkout, altitude chamber testing, and assembly.

Apollo 12 is the sixth Saturn V/Apollo space vehicle to be launched from Complex 39's Pad A, one of two octagonal launch pads which are 3,000 feet across. The major components of Complex 39 include:

1. The Vehicle Assembly Building, heart of the complex, is where the 363-foot-tall space vehicle is assembled and tested. It contains 129.5 million cubic feet of space, covers eight acres, is 716 feet long and 518 feet wide. Its high bay area, 525 feet high, contains four assembly and checkout bays and its low bay area - 210 feet high, 442 feet wide and 274 feet long - contains eight stage-preparation and checkout cells. There are 141 lifting devices in the building, ranging from one-ton hoists to two 250-ton high lift bridge cranes.

2. The Launch Control Center, a four-story structure adjacent and to the south of the Vehicle Assembly Building is a radical departure from the dome-shaped, "hardened" blockhouse at older launch sites. The Launch Control Center is the electronic "brain" of Complex 39 and was used for checkout and test operations while Apollo 12 was being assembled inside the Vehicle Assembly Building high bay. Three of the four firing rooms contain identical sets of control and monitoring equipment so that launch of one vehicle and checkout of others may continue simultaneously. Each firing room is associated with a ground computer facility to provide data links with the launch vehicle on its mobile launcher at the pad or inside the Vehicle Assembly Building.

3. <u>The Mobile Launcher</u>, 445 feet tall and weighing 12 million pounds, is a transportable launch base and umbilical tower for the space vehicle.

4. <u>The Transporters</u>, used to move mobile launchers into the Vehicle Assembly Building and then - with their space vehicles - to the launch pad, weigh six million pounds and are among the largest tracked vehicles known. The Transporters - there are two - are 131 feet long and 114 feet wide. Powered by electric motors driven by two 2,750-horsepower diesel engines, the vehicles move on four double-tracked crawlers, each 10 feet high and 40 feet long. Maximum speed is about one-mile-per-hour loaded and two miles-per-hour unloaded. The three and one-half mile trip to Pad A with a mobile launcher and space vehicle takes approximately seven hours. Apollo 12 rollout to the pad occurred on December 15, 1969.

5. <u>The Crawlerway</u> is the roadway for the transporter and is 131 feet wide divided by a median strip. This is the approximate width of an eight-lane turnpike and the roadbed is designed to accommodate a combined weight of more than 18 million pounds.

6. <u>The Mobile Service Structure</u> is a 402-foot-tall, 9.8 million pound tower used to service the Apollo space vehicle at the pad. Moved into place about the Saturn V/Apollo space vehicle and its mobile launcher by a transporter, it contains five work platforms and provides 360-degree platform access to the vehicle being prepared for launch. It is removed to a parking area about 11 hours before launch.

7. <u>A Water Deluge System</u> will provide about a million gallons of industrial water for cooling and fire prevention during the launch of Apollo 13. The water is used to cool the mobile launcher, the flame trench and the flame deflector above which the mobile launcher is positioned.

8. <u>The Flame Deflector</u> is an "A"-shaped, 1.3 million pound structure moved into the flame trench beneath the launcher prior to launch. It is covered with a refractory material designed to withstand the launch environment. The flame trench itself is 58 feet wide and approximately six feet above mean sea level at the base.

9. <u>The Pad Areas</u> - A and B - are octagonal in shape
and have center hardstands constructed of heavily reinforced
concrete. The top of Pad A stands about 48 feet above sea
level. Saturn V propellants - liquid oxygen, liquid hydrogen
and RP-1, the latter a high grade kerosene - are stored in
large tanks spaced near the pad perimeter and carried by pipe-
lines from the tanks to the pad, up the mobile launcher and into
the launch vehicle propellant tanks. Also located in the pad
area are pneumatic, high pressure gas, electrical, and industrial
water support facilities. Pad B, used for the launch of Apollo
10, is located 8,700 feet north of Pad A.

MISSION CONTROL CENTER

The Mission Control Center at the Manned Spacecraft Center, Houston, is the focal point for Apollo flight control activities. The center receives tracking and telemetry data from the Manned Space Flight Network which in turn is processed by the MCC Real-Time Computer Complex for display to flight controllers in the Mission Operations Control Room (MOCR) and adjacent staff support rooms.

Console positions in the two identical MOCRs in Mission Control Center fall into three basic operations groups: mission command and control, systems operations, and flight dynamics.

Positions in the command and control group are:

* Mission Director -- responsible for overall mission conduct.

* Flight Operations Director -- represents MSC management.

* Flight Director -- responsible for operational decisions and actions in the MOCR.

* Assistant Flight Director -- assists flight director and acts in his absence.

* Flight Activities Officer -- develops and coordinates flight plan.

* Department of Defense Representative -- coordinates and directs DOD mission support.

* Network Controller -- responsible to FD for Manned
Space Flight Network status and troubleshooting; MCC equipment operation.

* Surgeon -- monitors crew medical condition and informs
FD of any medical situation affecting mission.

* Spacecraft Communicator (Capcom) -- serves as voice
contact with flight crew.

* Experiments Officer -- coordinates operation and
control of onboard flight experiments.

* Public Affairs Officer -- reports mission progress
to public through commentary and relay of live air-to-ground
transmissions.

Systems Operations Group:

* Environmental, Electrical and Instrumentation
Engineer (EECOM) -- monitors and troubleshoots command/service
module environmental, electrical, and sequential systems.

* Guidance, Navigation and Control Engineer (GNC) --
monitors and troubleshoots CSM guidance, navigation, control,
and propulsion systems.

* LM Environmental and Electrical Engineer (TELCOM) --
LM counterpart to EECOM.

* LM Guidance, Navigation and Control Engineer (Control)--
LM counterpart to GNC.

* Booster Systems Engineer (BSE) (three positions) --
responsible for monitoring launch vehicle performance and for
sending function commands.

* Communications Systems Engineer (CSE) (call sign INCO)
and Operations and Procedures Officer (O&P) -- share responsibility for monitoring and troubleshooting spacecraft and
lunar surface communication systems and for coordinating MCC
procedures with other NASA centers and the network.

Flight Dynamics Group:

* Flight Dynamics Officer (FIDO) -- monitors powered
flight events and plans spacecraft maneuvers.

* Retrofire Officer (Retro) -- responsible for planning
deorbit maneuvers in Earth orbit and entry calculations on
lunar return trajectories.

* Guidance Officer (Guido) -- responsible for monitoring and updating CSM and LM guidance systems and for monitoring systems performance during powered flight.

Each MOCR operations group has a staff support room on the same floor in which detailed monitoring and analysis is conducted. Other supporting MCC areas include the space-flight Meteorological Room, the Space Environment (radiation) Console, Spacecraft Planning and Analysis (SPAN) Room for detailed spacecraft performance analysis, Recovery Operations Control Room and the Apollo Lunar Surface Experiment Package Support Room.

Located on the first floor of the MCC are the communications, command, and telemetry system (CCATS) for processing incoming data from the tracking network, and the real-time computer complex (RTCC) which converts flight data into displays useable to MOCR flight controllers.

MANNED SPACE FLIGHT NETWORK

The worldwide Manned Space Flight Network (MSFN) provides reliable, continuous, and instantaneous communications with the astronauts, launch vehicle, and spacecraft from liftoff to splashdown. Following the flight, the network will continue in support of the link between Earth and the Apollo experiments left on the lunar surface by the Apollo crew.

The MSFN is maintained and operated by the NASA Goddard Space Flight Center, Greenbelt, Md., under the direction of NASA's Office of Tracking and Data Acquisition. In the MSFN Operations Center (MSFNOC) at Goddard, the Network Director and his team of Operations Managers, with the assistance of a Network Support Team, keep the entire complex tuned for the mission support. Should Houston's mission control center be seriously impaired for an extended time, the Goddard Center becomes an emergency mission control center.

The MSFN employs 12 ground tracking stations equipped with 30- and 85-foot antennas, an instrumented tracking ship, and four instrumented aircraft. For Apollo 13, the network will be augmented by the 210-foot antenna systems at Goldstone, Calif. and at Parkes, Australia, (Australian Commonwealth Scientific and Industrial Research Organization).

NASA Communications Network (NASCOM). The tracking network is linked together by the NASA Communications Network. All information flows to and from MCC Houston and the Apollo spacecraft over this communications system.

The NASCOM consists of almost three million circuit miles of diversely routed communications channels. It uses satellites, submarine cables, land lines, microwave systems, and high frequency radio facilities for access links.

NASCOM control center is located at Goddard. Regional communication switching centers are in London, Madrid, Canberra, Australia; Honolulu and Guam.

TRACKING THE MOON

GOLDSTONE, CALIFORNIA

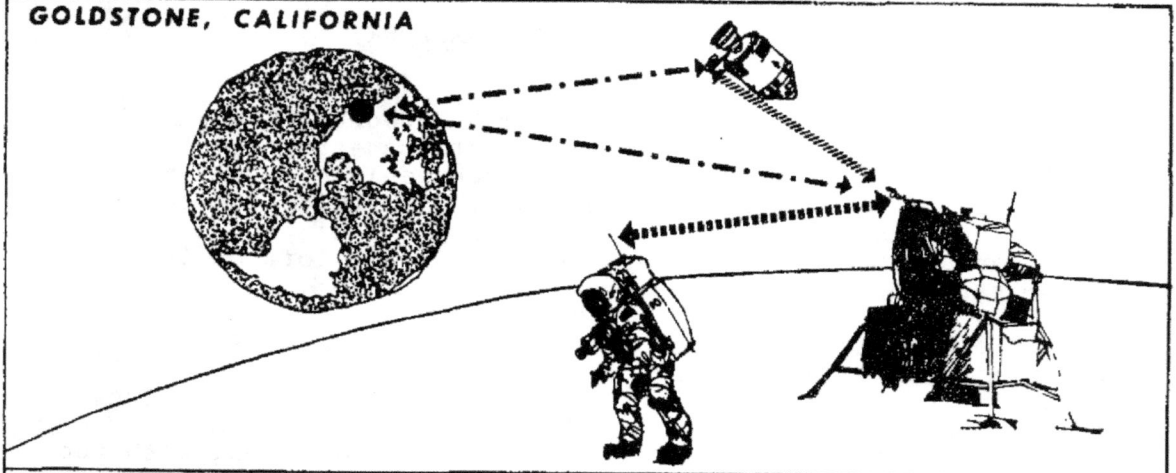

MADRID, SPAIN

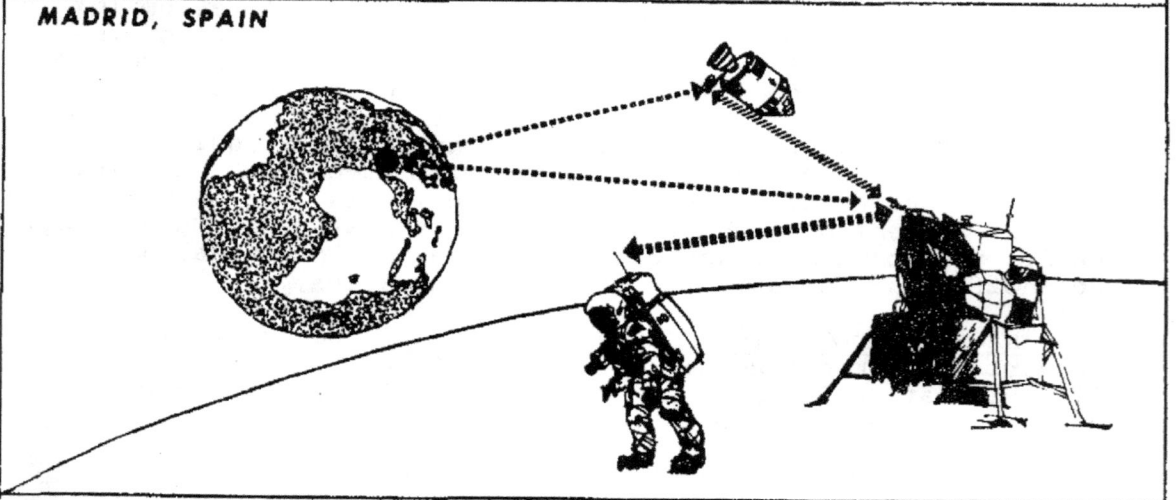

CANBERRA, AUSTRALIA

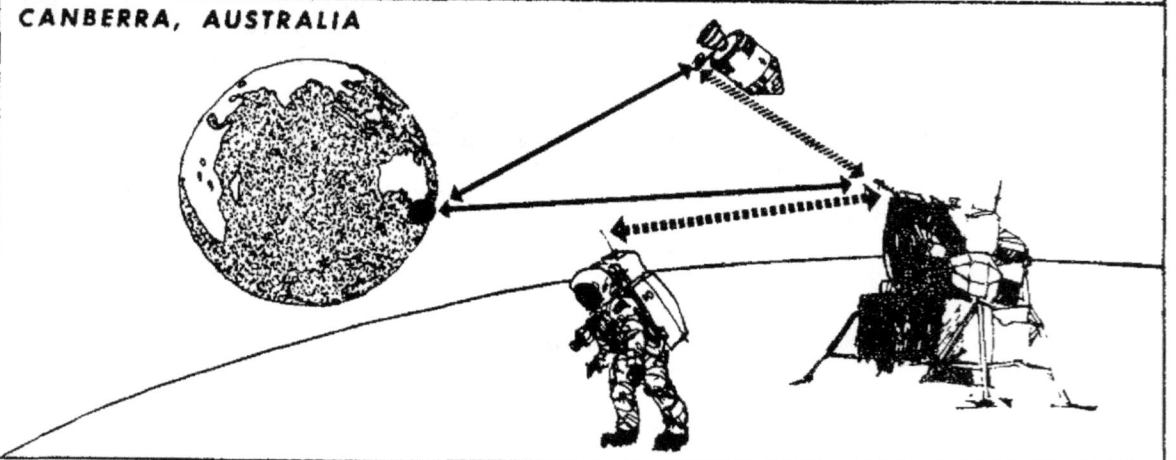

Three Intelsat communications satellites will be used for Apollo 13. One satellite over the Atlantic will link Goddard with stations at Madrid, Canary Islands, Ascension and the Vanguard tracking ship. Another Atlantic satellite will provide a direct link between Madrid and Goddard for TV signals received from the spacecraft. The third satellite over the mid-Pacific will link Carnarvon, Canberra, and Hawaii with Goddard through a ground station at Brewster Flats, Wash.

At Goddard, NASCOM switching computers simultaneously send the voice signals directly to the Houston flight controllers and the tracking and telemetry data to computer processing complexes at Houston and Goddard. The Goddard Real Time Computing Complex verifies performance of the tracking network and uses the collected tracking data to drive displays in the Goddard Operations Control Center.

Establishing the Link -- The Merritt Island tracking station monitors prelaunch test, the terminal countdown, and the first minutes of launch.

An Apollo instrumentation ship (USNS VANGUARD) fills the gaps beyond the range of land tracking stations. For Apollo 13 this ship will be stationed in the Atlantic to cover the insertion into Earth orbit. Apollo instrumented aircraft provide communications support to the land tracking stations during translunar injection and reentry and cover a selected abort area in the event of "no-go" decision after insertion into Earth orbit.

Lunar Bound - Approximately one hour after the spacecraft has been injected into its translunar trajectory (some 10,000 miles from the Earth), three prime tracking stations spaced nearly equidistant around the Earth will take over tracking and communicating with Apollo.

Each of the prime stations, located at Goldstone, Madrid and Canberra, has a dual system for use when tracking the command module in lunar orbit and the lunar module in separate flight paths or at rest on the Moon. These stations are equipped with 85-foot antennas.

The Return Trip -- To make an accurate reentry, data from the tracking stations are fed into the MCC computers to develop necessary information for the Apollo 13 crew.

MANNED SPACE FLIGHT TRACKING NETWORK

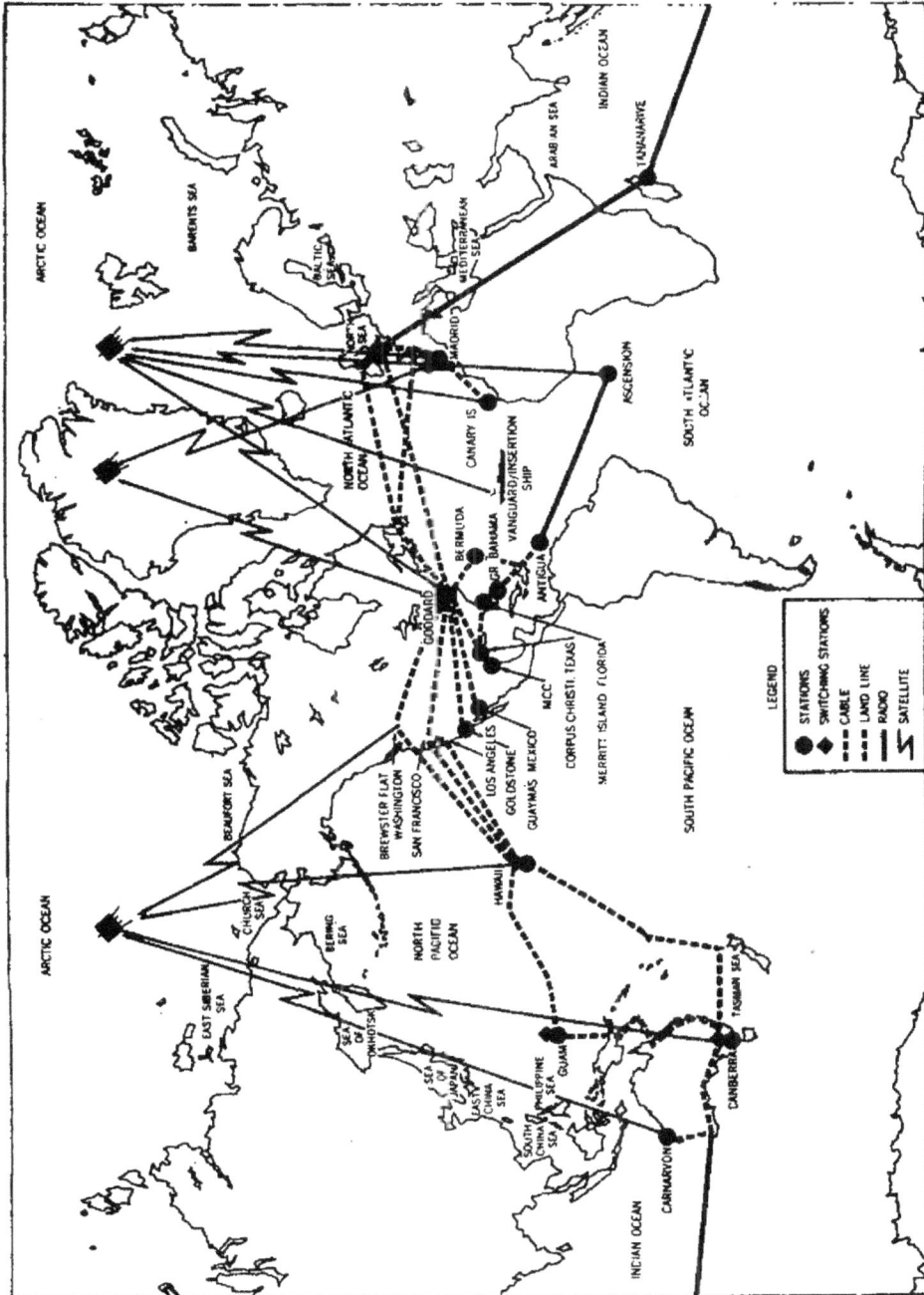

NASCOM

LEGEND

STATIONS
SWITCHING STATIONS
CABLE
LAND LINE
RADIO
SATELLITE

Appropriate MSFN stations, including the aircraft in the Pacific, provide support during the reentry.

Through the journey to the Moon and return, television will be received from the spacecraft at the three prime stations. In addition, a 210-foot antenna at Goldstone (an antenna of NASA's Deep Space Network) will augment the television coverage while Apollo 13 is near and on the Moon. For black and white TV, scan converters at the stations permit immediate transmission of commercial quality TV via NASCOM to Houston, where it will be released to U.S. TV networks.

Black and white TV can be released simultaneously in Europe and the Far East through the MSFN stations in Spain and Australia.

For color TV, the signal will be converted to commercial quality at the MSC Houston. A black and white version of the color signal can be released locally simultaneously through the stations in Spain and Australia.

Network Computers

At fraction-of-a-second intervals, the network's digital data processing systems, with NASA's Manned Spacecraft Center as the focal point, "talk" to each other or to the spacecraft. High-speed computers at the remote sites (tracking ship included) relay commands or "up-link" data on such matters as control of cabin pressure, orbital guidance commands, or "go-no-go" indications to perform certain functions.

When information originates from Houston, the computers refer to their pre-programmed information for validity before transmitting the required data to the spacecraft.

Such "up-link" information is communicated at a rate of about 1,200 bits-per-second. Communication of spacecraft data between remote ground sites and the Mission Control Center, via high-speed communications links, occurs at twice the rate. Houston reads information from these ground sites at 8,800 bits-per-second.

The computer systems perform many other functions, including:

Assuring the quality of the transmission lines by continually testing data paths.

Verifying accuracy of the messages.

Constantly updating the flight status.

For "down-link" data, sensors built into the spacecraft continually sample cabin temperature, pressure, and physical information on the astronauts such as heartbeat and respiration. These data are transmitted to the ground stations at 51.2 kilobits (12,800 decimal digits) per second.

At MCC the computers:

Detect and select changes or deviations, compare with their stored programs, and indicate the problem areas or pertinent data to the flight controllers;

Provide displays to mission personnel;

Assemble output data in proper formats;

Log data on magnetic tape for the flight controllers.

The Apollo Ship Vanguard

The USNS Vanguard will perform tracking, telemetry, and communication functions for the launch phase and Earth orbit insertion. Vanguard will be stationed about 1,000 miles southeast of Bermuda (28 degrees N., 49 degrees W.).

Apollo Range Instrumentation Aircraft (ARIA)

During the Apollo 13 TLI maneuver, two ARIA will record telemetry data from Apollo and relay voice communication between the astronauts and the Mission Control Center at Houston. The ARIA will be located between Australia and Hawaii.

For reentry, two ARIA will be deployed to the landing area to relay communications between Apollo and Mission Control at Houston and provide position information on the spacecraft after the blackout phase of reentry has passed.

The total ARIA fleet for Apollo missions consists of four EC-135A (Boeing 707) jets with 7-foot parabolic antennas installed in the nose section.

CONTAMINATION CONTROL PROGRAM

In 1966 an Interagency Committee on Back Contamination (ICBC) was established to assist NASA in developing a program to prevent contamination of the Earth from lunar materials following manned lunar exploration and to review and approve plans and procedures to prevent back contamination. Committee membership includes representatives from Public Health Service, Department of Agriculture, Department of the Interior, NASA, and the National Academy of Sciences.

The Apollo Back Contamination Program can be divided into three phases. The first phase covers procedures which are followed by the crew while in flight to reduce and, if possible, eliminate the return of lunar surface contaminations in the command module.

The second phase includes recovery, isolation, and transport of the crew, spacecraft, and lunar samples to the Manned Spacecraft Center. The third phase encompasses quarantine operations and preliminary sample analysis in the Lunar Receiving Laboratory.

A primary step in preventing back contamination is careful attention to spacecraft cleanliness following lunar surface operations. This includes use of special cleaning equipment, stowage provisions for lunar-exposed equipment, and crew procedures for proper "housekeeping."

Prior to reentering the LM after lunar surface exploration, the crewmen brush lunar surface dust or dirt from the space suit using special brushes. They will scrape their overboots on the LM footpad and while ascending the LM ladder, dislodge any clinging particles by a kicking action.

After entering and pressurizing the LM cabin, the crew doff their portable life support system, oxygen purge system, lunar boots, EVA gloves, etc.

Following LM rendezvous and docking with the CM, the CM tunnel will be pressurized and checks made to insure that an adequate pressurized seal has been made. During the period, some of the equipment may be vacuumed.

The lunar module cabin atmosphere will be circulated through the environmental control system suit circuit lithium hydroxide (LiOH) canister to filter particles from the atmosphere. A minimum of five hours weightless operation and filtering will essentially eliminate the original airborne particles.

The CM pilot will transfer lunar surface equipment stowage bags into the LM one at a time. The equipment transferred will be bagged before being transferred. The only equipment which will not be bagged at this time are the crewmen's space suits and flight logs.

Command Module Operations - Through the use of operational and housekeeping procedures the command module cabin will be purged of lunar surface and/or other particulate contamination prior to Earth reentry. These procedures start while the LM is docked with the CM and continue through reentry into the Earth's atmosphere.

During subsequent lunar orbital flight and the transearth phase, the command module atmosphere will be continually filtered through the environmental control system lithium hydroxide canister. This will remove essentially all airborne dust particles. After about 96 hours operation essentially none of the original contaminates will remain.

Lunar Mission Recovery Operations

Following landing and the attachment of the flotation collar to the command module, a swimmer will open the spacecraft hatch, pass in three clean flight coveralls and three filter masks and close the hatch.

Crew retrieval will be accomplished by helicopter to the carrier and subsequent crew transfer to the Mobile Quarantine Facility. The spacecraft will be retrieved by the aircraft carrier and isolated.

LUNAR RECEIVING LABORATORY (LRL)

The final phase of the back contamination program is completed in the MSC Lunar Receiving Laboratory. The crew and spacecraft are quarantined for a minimum of 21 days after completion of lunar EVA operations and are released based upon the completion of prescribed test requirements and results. The lunar sample will be quarantined for a period of 50 to 80 days depending upon results of extensive biological tests.

The LRL serves four basic purposes:

- Quarantine of crew and spacecraft, the containment of lunar and lunar-exposed materials, and quarantine testing to search for adverse effects of lunar material upon terrestrial life.

- The preservation and protection of the lunar samples.

- The performance of time critical investigations.

- The preliminary examination of returned samples to assist in an intelligent distribution of samples to principal investigators.

The LRL has the only vacuum system in the world with space gloves operated by a man leading directly into a vacuum chamber at pressures of about 10 billionth of an atmosphere. It has a low level counting facility, the background count is an order of magnitude better than other known counters. Additionally, it is a facility that can handle a large variety of biological specimens inside Class III biological cabinets designed to contain extremely hazardous pathogenic material.

The LRL covers 83,000 square feet of floor space and includes a Crew Reception Area (CRA), Vacuum Laboratory, Sample Laboratories (Physical and Bio-Science) and an administrative and support area. Special building systems are employed to maintain air flow into sample handling areas and the CRA, to sterilize liquid waste, and to incinerate contaminated air from the primary containment systems.

The biomedical laboratories provide for quarantine tests to determine the effect of lunar samples on terrestrial life. These tests are designed to provide data upon which to base the decision to release lunar material from quarantine.

Among the tests:

a.   Lunar material will be applied to 12 different
culture media and maintained under several environmental
conditions. The media will be observed for bacterial or
fungal growth. Detailed inventories of the microbial flora
of the spacecraft and crew have been maintained so that any
living material found in the sample testing can be compared
against this list of potential contaminants taken to the Moon
by the crew or spacecraft.

b.   Six types of human and animal tissue culture cell
lines will be maintained in the laboratory and together with
embryonated eggs are exposed to the lunar material. Based on
cellular and/or other changes, the presence of viral material
can be established so that special tests can be conducted to
identify and isolate the type of virus present.

c.   Thirty-three species of plants and seedlings will
be exposed to lunar material. Seed germination, growth of
plant cells or the health of seedlings are then observed,
and histological, microbiological and biochemical techniques
are used to determine the cause of any suspected abnormality.

d.   A number of lower animals will be exposed to lunar
material, including germ-free mice, fish, birds, oysters,
shrimp, cockroaches, houseflies, planaria, paramecia and
euglena. If abnormalities are noted, further tests will be
conducted to determine if the condition is transmissible from
one group to another.

The crew reception area provides biological containment
for the flight crew and 12 support personnel. The nominal
occupancy is about 14 days but the facility is designed and
equipped to operate for considerably longer.

Sterilization and Release of the Spacecraft

Postflight testing and inspection of the spacecraft is
presently limited to investigaiton of anomalies which happened
during the flight. Generally, this entails some specific
testing of the spacecraft and removal of certain components of
systems for further analysis. The timing of postflight testing
is important so that corrective action may be taken for subsequent flights.

The schedule calls for the spacecraft to be returned
to port where a team will deactivate pyrotechnics, and flush
and drain fluid systems (except water). This operation will
be confined to the exterior of the spacecraft. The spacecraft
will then be flown to the LRL and placed in a special room for
storage, sterilization, and postflight checkout.

LUNAR RECEIVING LABORATORY TENTATIVE SCHEDULE

April 20   Activate secondary barrier; support people enter
            Crew Reception Area and Central Status Station
            manned; LRL on mission status.

April 21   Command module landing, recovery.

April 22   First sample return container (SRC) arrives.

April 23   First SRC opened in vacuum lab, second SRC arrives;
            film, tapes, LM tape recorder begin decontamination;
            second SRC opened in Bioprep lab.

April 24   First sample to Radiation Counting Laboratory.

April 26   Core tube moves from vacuum lab to Physical-
            Chemical Lab.

April 26   MQF arrives; contingency sample goes to Physical-
            Chemical Lab; rock description begun in vacuum lab.

April 27   Biosample rocks move from vacuum lab to Bioprep
            Lab; core tube prepared for biosample.

April 28   Spacecraft arrives.

April 29   Biosample compounded, thin-section chips sterilized
            out to Thin-Section Lab, remaining samples from
            Bioprep Lab canned.

May 1   Thin-section preparation complete, biosample prep
            complete, transfer to Physical-Chemical Lab
            complete, Bioprep Lab cleanup complete.

May 3   Biological protocols, Physical-Chemical Lab rock
            description begin.

May 8   Crew released from CRA

May 26   Rock description complete, Preliminary Examination
            Team data from Radiation Counting Lab and Gas
            Analysis Lab complete.

May 28   PET data write-up and sample catalog preparation
            begin.

May 30   Data summary for Lunar Sample Analysis Planning
Team (LSAPT) complete.

June 1   LSAPT arrives.

June 2   LSAPT briefed on PET data, sample packaging begins.

June 6   Sample distribution plan complete, first batch
monopole samples canned.

June 8   Monopole experiment begins.

June 10   Initial release of Apollo 13 samples; spacecraft release.

June 14   Spacecraft equipment released

-more-

SCHEDULE FOR TRANSPORT OF SAMPLES, SPACECRAFT AND CREW

Samples

The first Apollo 13 sample return container (SRC) will be flown by helicopter from the deck of the USS Iwo Jima to Christmas Island, from where it will be flown by C-130 aircraft to Hawaii. The SRC, half the mission onboard film and any medical samples ready at the time of helicopter departure from the Iwo Jima, will be transferred to an ARIA (Apollo Range Instrumented Aircraft) at Hawaii for the flight to Ellington AFB, six miles north of the Manned Spacecraft Center, with an estimated time of arrival at 11:30 a.m. EST April 22.

The second SRC and remainder of onboard film and medical samples will follow a similar sequence of flights the following day and will arrive at Ellington AFB at an estimated time of 1 am EST April 23. The SRCs will be moved by auto from Ellington AFB to the Lunar Receiving Laboratory.

Spacecraft

The spacecraft should be aboard the Iwo Jima about two hours after crew recovery. The ship will arrive in Hawaii at 2 pm EST April 25 and the spacecraft will be offloaded and transferred after deactiviation to an aircraft for the flight to Ellington AFB, arriving April 28. The spacecraft will be trucked to the Lunar Receiving Laboratory where it will enter quarantine.

Crew

The flight crew is expected to enter the Mobile Quarantine Facility (MQF) on the Iwo Jima about 90 minutes after splash-down. Upon arrival at Hawaii, the MQF will be offloaded and placed aboard a C-141 aircraft for the flight to Ellington AFB, arriving at 1 am EST April 25. A transporter truck will move the MQF from Ellington AFB to the Lunar Receiving Laboratory-- about a two-hour trip.

APOLLO PROGRAM MANAGEMENT

The Apollo Program is the responsibility of the Office of Manned Space Flight (OMSF), National Aeronautics and Space Administration, Washington, D. C. Dale D. Myers is Associate Administrator for Manned Space Flight.

NASA Manned Spacecraft Center (MSC), Houston, is responsible for development of the Apollo spacecraft, flight crew training, and flight control. Dr. Robert R. Gilruth is Center Director.

NASA Marshall Space Flight Center (MSFC), Huntsville, Ala., is responsible for development of the Saturn launch vehicles. Dr. Eberhard Rees is Center Director.

NASA John F. Kennedy Space Center (KSC), Fla., is responsible for Apollo/Saturn launch operations. Dr. Kurt H. Debus is Center Director.

The NASA Office of Tracking and Data Acquisition (OTDA) directs the program of tracking and data flow on Apollo. Gerald M. Truszynski is Associate Administrator for Tracking and Data Acquisition.

NASA Goddard Space Flight Center (GSFC), Greenbelt, Md., manages the Manned Space Flight Network and Communications Network. Dr. John F. Clark is Center Director.

The Department of Defense is supporting NASA in Apollo 13 during launch, tracking and recovery operations. The Air Force Eastern Test Range is responsible for range activities during launch and down-range tracking. Recovery operations include the use of recovery ships and Navy and Air Force aircraft.

Apollo/Saturn Officials

NASA Headquarters

Dr. Rocco A. Petrone	Apollo Program Director, OMSF
Chester M. Lee (Capt., USN, Ret.)	Apollo Mission Director, OMSF
Col. Thomas H. McMullen (USAF)	Apollo Assistant Mission Director, OMSF
John D. Stevenson (Maj. Gen., USAF, Ret.)	Director of Mission Operations, OMSF
Maj. Gen. James W. Humphreys, Jr. (USAF, MC)	Director of Space Medicine, OMSF
John K. Holcomb,(Capt., USN, Ret.)	Director of Apollo Operations, OMSF
Lee R. Scherer,(Capt., USN, Ret.)	Director of Apollo Lunar Exploration, OMSF
James C. Bavely	Chief of Network Operations Branch, OTDA

Marshall Space Flight Center

Lee B. James	Director, Program Management
Dr. F. A. Speer	Manager, Mission Operations Office
Roy E. Godfrey	Manager, Saturn Program Office
Matthew W. Urlaub	Manager, S-IC Stage, Saturn Program Office
William F. LaHatte	Manager, S-II Stage, Saturn Program Office
Charles H. Meyers	Manager (Acting), S-IVB Stage, Saturn Program Office
Frederich Duerr	Manager, Instrument Unit, Saturn Program Office
William D. Brown	Manager, Engine Program Office

Kennedy Space Center

Walter J. Kapryan	Director of Launch Operations
Raymond L. Clark	Director of Technical Support
Edward R. Mathews	Apollo Program Manager
Dr. Hans F. Gruene	Director, Launch Vehicle Operations
John J. Williams	Director, Spacecraft Operations
Paul C. Donnelly	Launch Operations Manager
Isom A. Rigell	Deputy Director for Engineering

Manned Spacecraft Center

Col. James A. McDivitt, (USAF)	Manager, Apollo Spacecraft Program
Donald K. Slayton	Director, Flight Crew Operations
Sigurd A. Sjoberg	Director, Flight Operations
Milton L. Windler	Flight Director
Gerald Griffin	Flight Director
Glynn S. Lunney	Flight Director
Eugene F. Kranz	Flight Director
Dr. Charles A. Berry	Director, Medical Research and Operations

Goddard Space Flight Center

Ozro M. Covington	Director of Manned Flight Support
William P. Varson	Chief, Manned Flight Planning & Analysis Division
H. William Wood	Chief, Manned Flight Operations Division
Tecwyn Roberts	Chief, Manned Flight Engineering Division
L. R. Stelter	Chief, NASA Communications Division.

Department of Defense

Maj. Gen. David M. Jones, (USAF)   DOD Manager of Manned Space
Flight Support Operations,
Commander of USAF Eastern
Test Range

Rear Adm. Wm. S. Guest, (USN)   Deputy DOD Manager of Manned
Space Flight Support Operations,
Commander Task Force 140,
Atlantic Recovery Area

Rear Adm. Donald C. Davis, (USN)   Commander Task Force 130,
Pacific Recovery Area

Col. Kenneth J. Mask, (USAF)   Director of DOD Manned Space
Flight Support Office

Maj. Gen. Allison C. Brooks,   Commander Aerospace Rescue and
(USAF)   Recovery Service

-more-

Major Apollo/Saturn V Contractors

Contractor	Item
Bellcomm Washington, D. C.	Apollo Systems Engineering
The Boeing Co. Washington, D. C.	Technical Integration and Evaluation
General Electric-Apollo Systems Daytona Beach, Fla	Apollo Checkout, and Quality and Reliability
North American Rockwell Corp. Space Div., Downey, Calif.	Command and Service Modules
Grumman Aircraft Engineering Corp., Bethpage, N.Y.	Lunar Module
Massachusetts Institute of Technology, Cambridge, Mass.	Guidance & Navigation (Technical Management)
General Motors Corp., AC Electronics Div., Milwaukee, Wis.	Guidance & Navigation (Manufacturing)
TRW Inc. Systems Group Redondo Beach, Calif.	Trajectory Analysis LM Descent Engine LM Abort Guidance System
Avco Corp., Space Systems Div., Lowell, Mass.	Heat Shield Ablative Material
North American Rockwell Corp. Rocketdyne Div. Canoga Park, Calif.	J-2 Engines, F-1 Engines
The Boeing Co. New Orleans,	First Stage (SIC) of Saturn V Launch Vehicles, Saturn V Systems Engineering and Integration, Ground Support Equipment
North American Rockwell Corp. Space Div. Seal Beach, Calif.	Development and Production of Saturn V Second Stage (S-II)
McDonnell Douglas Astronautics Co., Huntington Beach, Calif.	Development and Production of Saturn V Third Stage (S-IVB)

International Business Machines Federal Systems Div. Huntsville, Ala.	Instrument Unit
Bendix Corp. Navigation and Control Div. Teterboro, N.J.	Guidance Components for Instrument Unit (Including ST-124M Stabilized Platform)
Federal Electric Corp.	Communications and Instrumentation Support, KSC
Bendix Field Engineering Corp.	Launch Operations/Complex Support, KSC
Catalytic-Dow	Facilities Engineering and Modifications, KSC
Hamilton Standard Division United Aircraft Corp. Windsor Locks, Conn.	Portable Life Support System; LM ECS
ILC Industries Dover, Del.	Space Suits
Radio Corp. of America Van Nuys, Calif.	110A Computer - Saturn Checkout
Sanders Associates Nashua, N.H.	Operational Display Systems Saturn
Brown Engineering Huntsville, Ala.	Discrete Controls
Reynolds, Smith and Hill Jacksonville, Fla.	Engineering Design of Mobile Launchers
Ingalls Iron Works Birmingham, Ala.	Mobile Launchers (ML) (Structural Work)
Smith/Ernst (Joint Venture) Tampa, Fla. Washington, D. C.	Electrical Mechanical Portion of MLs
Power Shovel, Inc. Marion, Ohio	Transporter
Hayes International Birmingham, Ala.	Mobile Launcher Service Arms
Bendix Aerospace Systems Ann Arbor, Mich.	Apollo Lunar Surface Experiments Package (ALSEP)
Aerojet-Gen. Corp. El Monte, Calif.	Service Propulsion System Engine

NASA-KSC APR/70

www.ingramcontent.com/pod-product-compliance
Lightning Source LLC
Chambersburg PA
CBHW051220200326

41519CB00025B/7182